Arming the Confederacy

Robert C. Whisonant

Arming the Confederacy

How Virginia's Minerals Forged
the Rebel War Machine

 Springer

Robert C. Whisonant
Geology
Radford University
Radford, VA, USA

ISBN 978-3-319-36241-0   ISBN 978-3-319-14508-2   (eBook)
DOI 10.1007/978-3-319-14508-2

Springer Cham Heidelberg New York Dordrecht London
© Springer International Publishing Switzerland 2015
Softcover reprint of the hardcover 1st edition 2015

Printed on acid-free paper

Springer International Publishing AG Switzerland is part of Springer Science+Business Media (www.springer.com)

Preface

I grew up in the Civil War. Actually, surrounded by it would be more accurate, because in Gaffney, South Carolina, in the 1940s and 1950s memories of the titanic struggle between North and South were everywhere. We lived almost directly across the street from the town cemetery where small United States banners adorned the graves of most of our war veterans. Others, however, had a different emblem—the Confederate battle flag. My Mother gave me a box of reunion commemorative medallions engraved with that same symbol her great uncle, an ex-Southern infantryman, had passed on to her. My stepfather had a sword allegedly brandished by an ancestor who served under fabled Rebel cavalryman Wade Hampton. Downtown at the intersection of Limestone and Buford Streets, an ever-faithful marble Confederate soldier, rifle at rest yet ready for action, kept watch over us all. These vessels of memory and many others fascinated me, and I came to love history, especially military, reading voraciously about armies of the past, most particularly those contending in the 1860s War of the Rebellion.

Although my interest in history never waned, geology coursework in college aroused a deeper passion because that way I could study not just the recent past but the long eons of Planet Earth's entire existence. After graduate school and a brief fling in the oil industry, I took a position teaching geology at Radford University, a mid-sized state school in the Appalachian Mountains of southwestern Virginia. Soon I realized that I had fortuitously landed in the heart of the foremost mineral-producing region of the entire Confederacy. Immense quantities of raw materials were mined and processed in the area, then carried away primarily by a vital railroad to supply distant weapons factories and armies.

Blue and Gray—North and South—clashed repeatedly during the war for control of the mineral works and railroad, fighting in some of the most difficult terrain in eastern North America. The topography itself, offering valleys as pathways for offensive movement and ridge tops for defensive stands, profoundly affected the course of the campaigns and battles. In these ancient mountains, mineral resources and landforms, key elements of the geological sciences, intimately intertwined with southwestern Virginia's Civil War history in ways almost unknown anywhere else. And that story had not yet really been told.

This book is my attempt to narrate that story. Other parts of Virginia are discussed, but the focus is on the southwestern quarter of the state. It was this region that supplied most of the Confederate salt, nearly all of the lead mined domestically, copious amounts of iron and niter, and even some coal, copper, and zinc. The bulk of these strategic materials traveled east on the Virginia and Tennessee Railroad to supply the center of the Southern armaments industries, the massive forges and foundries in Richmond, and the Confederacy's principal combat force, General Robert E. Lee's Army of Northern Virginia. From the earliest days of the hostilities, Union strategists viewed Southwest Virginia's minerals and railroad as high-value targets that must be eliminated. The engagements that resulted are a little known part of the Civil War but ones that deserve a much broader appreciation. Reaching that wider audience is my ultimate motive for writing this book.

I confess at the outset that I am a geologist, not a historian. Although this is a Civil War history book blended together with some geology, it is not a scholarly work intended for specialists in either field. I have done very little original research, relying instead on secondary sources that I believe authoritative. Confidence in the accuracy of these works is confirmed by my limited searches through primary archives, specifically the massive United States War Department *Official Records of the Rebellion* and published writings by soldiers who fought in the battles in Southwest Virginia. There are no reference footnotes in the text; instead, the Notes section at the end of each chapter lists the works used for that material and also specifies the sources of the direct quotes. The quotes are referenced to the original archives if I had access to those documents. In most cases, however, the secondary sources where I found the quotes are cited. A complete Bibliography of all references used to prepare the book follows the chapters.

The book is based on a series of articles first written for the Virginia Division of Mineral Resources about the geological aspects of the war in Southwest. I have, however, greatly revised and expanded those pieces for this volume. The scope has been broadened to look at the history of the mineral operations elsewhere in Virginia and to examine how the products manufactured from them—things such as gunpowder, bullets, and heavy artillery—undergirded the Southern war effort. In order to provide more context for understanding why minerals held such importance in the Civil War, fresh information is included on the martial use from ancient times to the 1860s of five raw materials—niter, lead, salt, iron, and coal. All were essential to both belligerents for making weapons, powering the machines of battle, or sustaining the soldiers. Each of these five extracted natural resources came from southwestern Virginia in significant quantities and without them the South's capacity to prosecute the war would have been seriously diminished. In addition to the discussions of the minerals, new material on terrain and railroads is included because of their fundamental impact on military events in western Virginia and the rebellious areas beyond.

Radford, VA, USA   Robert C. Whisonant

Acknowledgements

The inspiration for writing this book came from a number of people offering comments after hearing my public lectures on the role played in the Civil War by Virginia's minerals and the Virginia and Tennessee Railroad. "That was so interesting" they would say. "You should write a book." And so I have. Other books already exist that deal with the historical aspects of the fights for the minerals and the railroad that hauled them, but this book attempts to bring in geological and topographical information that adds context to the Civil War military events in Southwest Virginia and beyond.

I am indebted to many people for the creation of this book. First of all, I wish to thank Radford University and the Department of Geology for the unwavering support they have provided over my years of active teaching and now in my appointment as Professor Emeritus and Research Associate. My present colleagues in Geology—Drs. Jon Tso (Chair), Steve Lenhart, Chester (Skip) Watts, Parvinder Sethi, Beth McClellan, and Mr. George (Paki) Stephenson—continue to offer their "rock" solid support of my many endeavors. Two other Radford University scientific colleagues, archaeologist Cliff Boyd and geophysicist Rhett Herman, keep me busy and engaged with various Civil War battlefield protection and preservation projects. I also am grateful to a recent Radford University graduate in Geology, Chris Bolgiano, for drafting the maps in this book.

The inception of this volume goes back to the mid-1990s when I began researching and writing a set of articles on geology and the Civil War in southwestern Virginia for the Virginia Department of Mineral Resources (VDMR). At that time, Stan Johnson and Gene Rader at the VDMR in Charlottesville encouraged these efforts and guided them to publication in the *Virginia Minerals* series. The first of the topics addressed was the Wythe County lead mines, which brought me into contact with Cecil "Pete" Spraker, a historian at the New River Trail State Park. Pete spent many hours with me at the Shot Tower and old lead mines area, generously sharing his knowledge as we spoke and explored. Mr. George Mattis, a librarian at Wytheville Community College, helped me get started in those days as well.

The next piece in the *Virginia Minerals* series concerned the salt works at Saltville, and here I must thank Charlie Bill Totten and Harry Haynes. These two

Saltville residents have deep knowledge of Saltville's extraordinary history and they graciously let me pick their brains, then and now. A long telephone conversation with author William Marvel also contributed mightily to the Saltville research. My studies for the VDMR series then turned to the May 1864 railroad raid into the New River Valley. A discussion with author Howard McManus on his book about that campaign assisted this work tremendously. Radford University historian Linda Killen helped greatly as she took me around to the relevant sites in the region.

Follow-up work to produce a paper on the iron industry in southwestern Virginia led me back to Peter Spraker and George Mattis for material on this topic. Pete showed me several old iron furnaces still standing in this area. For an article on the Montgomery County coal mines, I am grateful to Robert Freis, an advocate for the preservation of the locale's mining heritage, and Radford University professor Dr. Mary La Lone for sharing information about the old Merrimac mines and other coal operations in the vicinity. The Civil War articles for the *Virginia Minerals* series ended with the story of the niter caves. Dave Hubbard at the VDMR and Dr. Ernst Kastning, then a geology professor at Radford University, provided publications and personal information concerning many aspects of saltpeter cave geology and the history of mining the niter deposits.

Since the completion of the VDMR work and with this book in mind, I have broadened the research to include other parts of Virginia and additional states in the Confederacy. At the same time, I extended my studies into the military use of the minerals in human history and especially during the Civil War. This has involved much time spent reading other authors and experts too numerous to mention. Here I must thank the librarians at Radford University in the special collections and inter-library loans sections who retrieved for me many excellent sources of information. I am especially grateful to Dr. Ernst H. Kastning, Jr. for the cave photographs in Chap. 5 and to Colonel Lewis Jeffries and Scott Gardner for providing the historic locomotive photograph in Chap. 13.

The reviewers of the manuscript deserve special mention. Dr. Judi Ehlen, a research colleague and seasoned editor of books concerned with military geology and geography, provided an in-depth critique. Most of her comments and sugges-tions have been used, and the book is much the better for it. Dr. Cliff Boyd, archae-ology professor at Radford University, and Dr. Bill Grant, economics professor at James Madison University, provided extremely valuable input from the perspective of disciplines outside of geology and history. Brenda Whisonant, a fine writer her-self, gave helpful comments on the first draft. Of course, any errors of commission or omission are mine entirely. I also wish to thank Ron Doering, a Springer Publishing Company editor in New York, for his patience in shepherding me through the ins and outs of taking my work from manuscript to final book form.

Most of all, I owe my deepest gratitude to my loving wife Brenda, daughter Dell, and son Bob. I suppose nearly all families that include a geologist can be described as "long suffering." We rock hounds can be difficult to live with, as when we are barreling down the highway with one eye on the road (at least some of the time) and the other on the rock cuts flitting past. Too often, we are compelled to stop, examine, sample, and photograph the exposures on *both* sides of the road. All of this and much more my family has borne with good grace and uncomplainingly (most of the time). So thank you, my Dears, for putting up with my wanderings to obscure places and into offbeat interests, but most of all for the meaning and richness you have brought into my life—this book is dedicated to you.

Contents

About the Author

Robert C. Whisonant is a native of Gaffney, South Carolina, and holds degrees in geology from Clemson University (B.S., 1963) and Florida State University (M.S., 1965, Ph.D., 1967). Dr. Whisonant worked for Exxon Inc. as a petroleum geologist from 1967 to 1971, then taught and served as an administrator at Radford University for over 31 years. He retired in 2002 and presently is Professor Emeritus and Research Faculty at the university.

Dr. Whisonant has authored or co-authored over 100 professional publications and consulted on a number of engineering geology projects. Because of his numerous teaching and professional contributions, Dr. Whisonant won the 1993 Neil Miner Award given by the National Association of Geology Teachers for outstanding university-level teaching in the United States. He also received Radford University's Professorial Excellence Award in 2000. In 2002, he was given the Outstanding Faculty Award by the State Council of Higher Education in Virginia.

Dr. Whisonant's current research includes developing the connections between geology and the American Civil War campaigns and battles. He has studied Virginia's mineral deposits (salt, lead, iron, coal, and niter) and their crucial role in supporting the Confederate war machine. He has analyzed the effects of geology and topography on casualty rates in national battlefields such as Gettysburg, Chancellorsville, and Antietam. He has published extensively on these topics and presented his research to audiences such as the Geological Society of America and the International Conference on Military Geography and Geology at the United States Military Academy in West Point, New York, in 2003, Nottingham, England, in 2005, and Quebec City, Canada, in 2007. He also directed projects involving mapping and preservation of the Civil War battlefield features at Saltville, Virginia, where the work was supported by NASA and the National Park Service (American Battlefield Protection Program).

Chapter 1
Introduction

"Victory had always gone to the side with the greatest material resources"

Nowhere is the impact of mineral resources on the military history of western civilization better shown than by the story of the Battle of Salamis in 480 BCE, the culmination of the struggle between the Greeks and Persians for eastern Mediterranean hegemony. After blunting an earlier Persian threat at Marathon, the citizens of Athens were advised by the Oracle at Delphi that a "wooden wall" would save the city. The Athenians came to believe that this meant building a much larger fleet of ships and decided to use the revenue from the silver mines at nearby Laurion to accomplish the task. Silver from Laurion had long given Athens the financial means to support its rise to prominence in the Aegean world. Now, the citizens voted to forego their usual dividends from the mines, instead using the money to add 200 ships to their naval armada.

At Salamis, the enhanced Greek flotilla delivered a crushing defeat to the Persian foe. Xerxes, king and leader of the Asian invaders, watched in horror from a throne on a hilltop overlooking the sea as the more nimble Greek galleys smashed his ponderous vessels to pieces. The Persians withdrew, having lost perhaps their best chance to permanently choke off rising Hellenic military power. Chroniclers have long rated Salamis as one of the most decisive engagements in world history—the battle that saved the democratic and scientific foundations of the West.

Successful nations are built on wealth. Throughout human history, mineral resources[1] have provided the financial means and raw materials that powered the rise of empires, granting them the capacity to wage war so that wealth could be

---

[1] Minerals are defined by geologists as naturally occurring inorganic materials with a fixed and symmetrical internal structure that gives them important properties such as crystal faces (so prized by gem collectors and jewelry wearers), hardness, strength, and the like. In this book, we will consider mineral resources simply as natural substances that can be economically extracted from the earth.

© Springer International Publishing Switzerland 2015
R.C. Whisonant, *Arming the Confederacy*, DOI 10.1007/978-3-319-14508-2_1

maintained and expanded. Historian Paul Kennedy, referring to struggles involving Great Power countries over the last five centuries, put it succinctly: "…victory has always gone to the side with the greatest material resources" [106]. For ancient Egypt, gold was the foundation of dominion. Gold and silver supported the economies and armies of classical Greece and Rome, and centuries later the flow of those precious metals from the Americas enabled Spain to gain supremacy. More recently, iron and coal, the basic drivers of the Industrial Revolution, brought stunning economic and political benefits first to England, then to the United States. Each of these countries achieved international financial and military dominance built at least partly on an abundance of such material resources. Oil now reigns as the new natural assets master, and global wealth and economic influence fall to those nations who can access and utilize the most black gold.

In addition to furnishing much of a country's financial base, mineral products are required to manufacture the physical means for engaging in armed conflict with its enemies. Warring societies early on put copper to use as a weapon, then bronze, and ultimately iron, which has remained an armaments mainstay for millennia. Over the centuries, a host of other natural materials have been indispensable for fighting— salt for the nutrition of the soldiers and army animals; niter (or saltpeter) for gunpowder; lead for bullets; and coal, oil, and now uranium to fuel the war industries and machines of the world's armies, navies, and air forces.

The wars of the twentieth century between modern machine-dependent martial forces exemplify just how essential adequate supplies of a key mineral resource— petroleum—have become in conflicts. British foreign secretary George Curzon remarked after World War I that "…the Allies floated to victory on a wave of oil" [279]. World War II was in many ways even more of an oil war. It began for the United States when this nation cut off Japan's supply and the Imperial Navy retaliated at Pearl Harbor. When American submarines at last shut off the shipping bringing petroleum to the home islands, Japan's hopes of triumph sank with the torpedoed tankers. Germany grew increasingly desperate for oil during the war, and much of the fighting in Russia stemmed from the Nazi thrust to capture the oil-rich Caucasus area. As German petroleum supplies dwindled, its chemists and engineers finally turned to obtaining oil from coal. Although innovative (parts of this technology remain in use today), this effort could not come close to adequately feeding the Reich's war machines. Lord Curzon's comment from 1918 rang still truer in 1945.

The American Civil War matched two industrialized opponents, but one side had a preponderant share of raw materials, manufacturing ability, and population that enabled it to emerge victorious. On the eve of battle in 1860, 90 per cent of the manufacturing capacity resided with the Union. Its factories made 97 per cent of the firearms, 94 per cent of the cloth, 93 per cent of the pig iron, and over 90 per cent of the boots and shoes. The disparity in the capability to make items necessary to the conduct of hostilities was based primarily on the North's possession of more mineral riches, most notably iron and coal. Moreover, the Northern states had twice the railroad track mileage as the South, and over those rails traveled coal-burning trains.

The rise of mechanized manufacturing and transportation north of the Potomac River had meanwhile attracted throngs of immigrants, many of them skilled in the trades and crafts useful to businesses. These recently arrived citizens made up a considerable part of a Federal population that was twice that of the Confederacy. Yet despite these advantages, as historian James McPherson has pointed out, "… numbers and resources do not prevail in war without the will and skill to use them" [148]. The North endured discouraging defeats early in the conflict, but in due time United States president Abraham Lincoln and his generals—Ulysses Grant, William Sherman, and Philip Sheridan in particular—came to wage "total war," attacking the enemy's infrastructure, industry, agriculture, mines, and foundries as well as his armies in the field.

The South in fact was endowed with considerable assets of its own, including a people determined to contest every bit of a very large land mass, some heavy industry and weapons manufacturing (located mostly in Virginia), over 9,000 miles of railroad track, a globally important agricultural commodity in cotton and slaves to work the fields, and innovative military leadership, at least in the Eastern Theater. Furthermore, the Confederacy possessed extensive mineral deposits, although mining and refining of these were typically underdeveloped compared to the North and the transportation system often too deficient to deliver them.

In terms of mining activity, no Confederate state surpassed Virginia where large amounts of lead, salt, iron, niter, and coal were located. The southwestern quarter of the state produced most of the Confederacy's lead and salt, along with significant quantities of the other minerals. Still, these natural resources held no value unless they could be moved to where they were needed. That task fell primarily to the Virginia and Tennessee Railroad. Over these steel rails moved the mineral products as well agricultural output vital to keeping Confederate armies in the field. Union Colonel Rutherford B. Hayes, involved in many of the military actions in Southwest Virginia, did not exaggerate when he called the rail line "…the jugular vein of the Confederacy" [165].

Almost from the beginning, Federal military strategists concerned with southwestern Virginia locked onto three prime targets: the salt works at Saltville, the Virginia and Tennessee Railroad running from Lynchburg to Bristol, and the Wythe County lead mines along the New River at Austinville (Figs. 1.1 and 1.2). In the first years of the contest, Saltville, deservedly named for its chief product, drew the most Union attention. Later on, however, the lead works became the chief objective; had the North fully understood the critical nature of the South's lead supply, the order of priority might have been reversed much earlier.

Saltville was important to be sure, but other salt works functioned, particularly in Alabama and at some coastal locations, and new ones could have been started up. The Virginia and Tennessee tracks stretched over 200 miles and could never have been completely destroyed. Until the last days of the war, damage from localized Union attacks on weak points was quickly repaired; in 1864 workmen rebuilt the burned New River railroad bridge in only a few weeks. The Wythe County lead works were different. Here stood the only large-scale mining and smelting facilities for that metal within the Confederate states. Elsewhere there existed just a handful of sporadically active works, none anywhere near the capacity of the Wythe County installations, and fresh lead deposits simply were not to be had. If the lead mines along the New River went down, the South would have been hard-pressed to keep bullets flying from the armies' rifles.

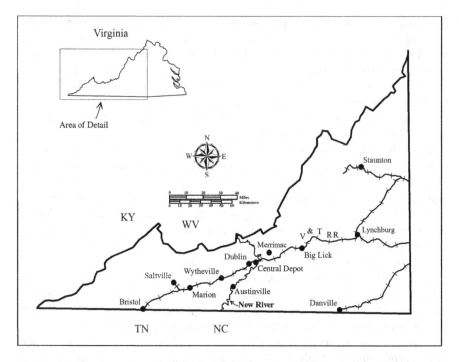

Fig. 1.1 Location map of southwestern Virginia showing key Civil War features. Note the three main targets of Union attacks in the region: Saltville (salt works), Austinville (lead mines), and Central Depot (New River Railroad Bridge). Central and Big Lick are today the cities of Radford and Roanoke, respectively. "V&T RR" is the Virginia and Tennessee Railroad

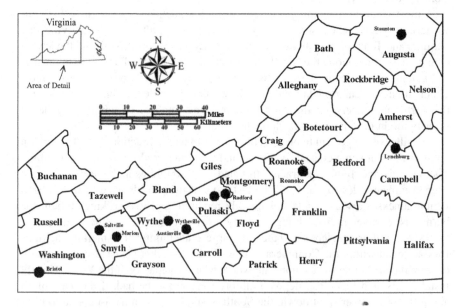

Fig. 1.2 Location map of counties and towns in southwestern Virginia

For the first two years, Union raids into Southwest Virginia were fairly small-scale efforts. Sorely needed men and supplies went to other theaters where big, fearfully bloody campaigns and battles consumed human and material resources at rates no one had anticipated. Confederate victories in the Eastern Theater threatened the very survival of the Union. After the battles of Gettysburg and Vicksburg in July 1863, things changed. Now the decision began to swing in favor of the North, already rapidly gearing up for the new kind of industrialized war both sides found themselves prosecuting. As more troops and equipment became available, Northern penetrations into the mountains of southwestern Virginia occurred more frequently, involving more substantial forces.

Some people, perhaps, would not consider the mineral and railroad actions to be true battles, at least not measured against huge engagements like Shiloh, Chancellorsville, and Gettysburg. At such major clashes, tens of thousands of soldiers on both sides took part and the span of fighting usually lasted more than a day. The greatest battle in southwestern Virginia occurred at Cloyds Mountain, where no fewer than 10,000 troops fought for about an hour. Nevertheless, the high casualty rates in such a short length of time attest to the same ferocity as in the much bigger contests. In only 52 min, over 1,200 soldiers—10 per cent of the Union troops and nearly a quarter of the Confederates—lost their lives or fell wounded.

Furthermore, in the mountains of southwestern Virginia as well as eastern Kentucky and Tennessee, combat could be unusually vicious. Many of the men facing each other here were locals who knew one another or were even blood relatives. Settling old grudges violently, a common way of life in the hill country, sometimes turned Civil War battles into opportunities to seek personal vengeance as well as simply to kill men invading their homeland.

The October 1864 Battle of Saltville illustrates the brutality of warfare in the southern Appalachian Mountains. An assemblage of Confederate army regulars and local partisans defeated a larger Union army from Kentucky, which included a regiment of black soldiers. The Northerners retreated in haste during the night, leaving many of their wounded and dead lying on the field where they fell. Over the next several days, a band of Southern fighters led by the notorious outlaw Champ Ferguson murdered the wounded, both black and white, left on the battlefield or receiving treatment in a nearby temporary hospital. At the hospital, Ferguson looked for and found a white lieutenant named Elza Smith. Stating that "…I have a begrudge against Smith…" [135], Ferguson shot the Federal officer to death where he lay. Known as the Saltville Massacre, these slayings are among the worst atrocities of the Civil War.

The central theme of this book is an exploration of the contributions to the Confederate war effort made by Virginia's mineral resources, especially those found in the remote mountains of Southwest Virginia. The campaigns and battles touched off by the extraordinary significance of the area's mineral works as well as the Virginia and Tennessee Railroad remain obscure and unheralded to this day. And yet, that significance reached far beyond the region's ancient hills and valleys, enabling the Southern armies to keep up the struggle for four long years.

The story told in these pages is a mix of history and geology. The following three chapters serve as context for the two main geological factors—minerals and terrain—woven throughout the historical narrative. Chapter 2 looks at the value of mineral resources in warfare from earliest human history into Civil War times and the rise of a strong mineral industry in pre-war America. These natural resources and the manufacturing facilities based upon them helped immensely to drive the young United States toward global economic and political prominence by the 1860s. Having adequate supplies of such mineral assets would be fundamental to the prosecution of the war for both Union and Confederacy.

Chapters 3 and 4 explain the profound effects of terrain, the physical shape of the earth's surface, on early American history and Civil War events in particular. The rocks, soils, and climate of the eastern United States had much to do with the emergence of two sections with distinctly different (and ultimately antagonistic) cultures, economies, and political views that brought them to blows in 1861. In the war years, terrain—the one constancy of warfare "...that forever imposes its realities on the course of battle" [283]—affected the campaigns and battles in all theaters. In southwestern Virginia, the rugged mountains and intervening valleys provided both opportunity and challenge for the contending sides, albeit considerably more challenge than opportunity for the attacking North.

These contextual chapters are followed by discussions of five mineral resources required in large quantities to wage war in the 1860s: niter, lead, salt, iron, and coal. The history of the military use of each is sketched, with emphasis on the Confederacy's determined efforts to extract, process, distribute, and utilize the natural materials. The military engagements in Southwest Virginia where the Southern mines and processing facilities were concentrated are the direct result of the Union's continuing attempts to thwart those efforts and eliminate these strategic assets.

The last two chapters of the book deal with the principal carrier of the mineral resources and a major mover of farm products, the Virginia and Tennessee Railroad. Chapter 13 describes the importance in the Civil War of railroads in general and Virginia's network, the largest in the Southern states, specifically. Within that system lay the Virginia and Tennessee, destined to become the "...Confederacy's single most important east-west connector..." and "...the chief supply line for Confederate forces in Virginia" [166]. Chapter 14 recounts the May 1864 attack against the Virginia and Tennessee ordered by General Ulysses S. Grant himself. This was the largest of all the Federal raids into southwestern Virginia, and from it came the short but exceptionally fierce Battle of Cloyds Mountain and the dramatic artillery duel at the New River railroad bridge. An Epilogue recounting the post-war fates of the mineral industries, railroad, and people involved in the fighting in Southwest Virginia concludes the chapters of the book.

Such was the war against Southwest Virginia's mineral manufactories and rail line that were so crucial to the Confederacy's survival. When it was all over, did the episodes of bravery and butchery in these mountains make any difference? I think they did. The North certainly believed that shutting down the mineral works and the railroad here would seriously degrade the South's ability to keep fighting. Given command of all the Federal armies in March 1864, Grant immediately ordered a major advance against the Virginia and Tennessee Railroad, stating that cutting the tracks and obliterating the bridge over the New River were the most important tasks the Union expedition could accomplish. After peace came, Confederate ordnance and supply officers commented on the inestimable worth of Virginia's lead and salt to the Southern states. In the end, though considerable damage had been inflicted by the relentless Union incursions, the salt, lead, and other mineral works were never completely demolished, turning out their products for a dying Confederacy to the very last days. The railroad served to the end as well. Although miles of track were torn up, the trains still ran locally in many areas at the close of the war.

Should the North have tried harder to completely eradicate the mineral works and the Virginia and Tennessee? Militating against such efforts were the issues of feasibility and resources. From the outset, the Union high command never seemed to fully appreciate the physical difficulty of invading southwestern Virginia. One Federal commander admitted after the war that campaigning in the southern Appalachians seemed so easy just looking at maps on flat pieces of paper. In reality, the minerals and railroad resided deep amid thickly timbered, tall mountain ranges with few roads for access. For most of the war, getting at them required the Union forces to approach from the west directly across the trend of the Appalachians, fighting their way in and out, and doing battle on steep terrain that greatly favored the defenders. To achieve this successfully would indeed have taken a substantial commitment of precious troops and equipment not readily available until late in the war. And yet, given the profound importance of the lead, salt, and railroad, a more determined effort against these assets should have been mounted much earlier. Failure to do so was perhaps the biggest strategic mistake made by the Federals in dealing with Southwest Virginia.

It is difficult to imagine how the South could have contended as long as it did had the lead- and salt-making facilities and the Virginia and Tennessee been captured or disabled early in the conflict. Making up the loss of the great majority of its salt and nearly all of the domestic lead would have been a daunting challenge for the Southern cause, and rebuilding wrecked stretches of a 200-mile-long railroad virtually impossible. The ruination of these resources in all likelihood would have shortened the conflagration to some extent, possibly by a year or even more. In a real sense, Southwest Virginia's minerals and railroad played a leading role in forging and maintaining the Confederate war machine.

Notes

The introductory story of how the silver mines near Athens made Greek victory possible at the Battle of Salamis is taken from Mirsky and Bland [156] and Craig et al. [36]. The emphasis on the importance in warfare of material resources, many of which are mineral-based, I found in Kennedy [108]. Yergin [280] is the source of the necessity for oil in World Wars I and II. The overview of the Civil War as a conflict in which material resources played a major part is abstracted from Kennedy [108] and McPherson [149]. The discussion of the significance of Virginia's mineral contributions in the Civil War is from Boyle [22], Schroeder-Lein [203], and Lynch [127]. Noe [169] provided the information on the Virginia and Tennessee Railroad. The brief look at the Civil War military events in southwestern Virginia is summarized from Walker [246], Johnson [98], McManus [144], Marvel [131], and Mays [134, 136]. The information in the chapter by chapter outline of the book is an amalgamation of all the sources used in the book and listed in the Bibliography. Of course, the opinions expressed at the end of the Introduction are strictly my own, crystallized over the last 20 years of gathering information and thinking about the role of Southwest Virginia's minerals and railroad in the Civil War.

Chapter 2
Minerals and Warfare

"Dig, mine, and search for all Manner of Mines of Gold, Silver, and Copper"

Civil War troops marching off to battle in 1861 most likely gave no thought to the origin of the weapons they carried or the provisions in their haversacks. But where did the iron come from to make the rifles they shouldered and the cannon, shot, and shell trailing along behind them? And from where was the coal mined that fired the forges and foundries to cast all that ordnance? Whence came the lead bullets in their pouches or the gunpowder that propelled those deadly projectiles or the salt that preserved the meat in their rations? The answer to all these questions is the same—the mineral industries of the North and South. The lead and salt for the Confederate soldiers in all likelihood were taken from extensive deposits in southwestern Virginia, by far the leading source of these commodities in the South. The coal and gunpowder niter probably originated in Virginia as well, which provided more of these mineral products than any other Southern state. Virginia contributed a substantial amount of iron as well, ranking a close second to Alabama, the leading source.

The Civil War has been called "...the first real industrialized 'total war' on proto-twentieth century lines..." [107] in the sense that it pitted against each other two segments of an advanced nation. The Northern states had a larger population, greater financial wealth, a better-developed infrastructure, and a far stronger manufacturing base rooted in bountiful natural resources, particularly iron and coal. Still, the Union was not irrevocably predestined to win; it truly was a "near run thing," as the Duke of Wellington, victor at Waterloo, said of that battle. Had the South triumphed at Antietam in September 1862, recognition by England and France as an independent nation and a negotiated settlement with the North might have followed. If Atlanta had not fallen to General Sherman in August 1864, President Lincoln might have lost the election that fall and his government replaced by one willing to discuss peace terms with a separate Confederate States of America.

© Springer International Publishing Switzerland 2015
R.C. Whisonant, *Arming the Confederacy*, DOI 10.1007/978-3-319-14508-2_2

Fig. 2.1 Confederate railroad yard in Nashville, Tennessee, in 1864. The railroads used huge quantities of iron, but the Confederacy could not produce enough to meet the surging demand during the Civil War. Photograph courtesy of the Library of Congress

None of that happened, of course, and the Civil War dragged on, a long, wearing conflict that relentlessly tested the economic and industrial strength, as well as the armies, of the protagonists. As things turned out, the Confederacy could not keep up in this lengthy duel that consumed men and resources in unprecedented proportions, and the United States won.

Iron for the railroads is a dramatic example of how the lack of a critical mineral product profoundly impacted the Confederate ability to conduct war (Fig. 2.1). The value of the steel rails became evident at the first big clash between North and South at Manassas in July 1861. This engagement clearly "…demonstrated the importance of railroads for moving large numbers of men to strategic positions…" [77] when the delivery by Rebel trains of additional troops just before the battle turned the tide. Time and again over the next 4 years, railroads transported soldiers, supplies and equipment, artillery, foodstuffs, and raw materials to support the military efforts of the belligerents. Tremendous quantities of iron would be needed to keep the railways running. The Confederacy's lines alone required an estimated 50,000 tons each year for repairs and new construction. Yet the combined output of all the Southern furnaces and forges amounted to only 20,000 tons, and much of that had to go toward weapons, armor plate for the ironclad naval vessels, and a variety of other military as well as civilian needs.

Other wartime problems plagued the Confederate iron industry. Persistent labor shortages on both the supply and production sides held back output. The charcoal-burning furnaces making the raw pig iron from ore in the mining areas consumed copious amounts of wood and the rapid cutting away of the timber stands as the hostilities intensified created a scarcity of that fuel. The underdeveloped road and railroad network made getting the pig iron to the distant steel mills inordinately difficult. Union advances brought about debilitating loss of territories where the raw iron was generated.

All of this resulted in a chronic shortage of iron throughout the struggle that kept the Confederate railways from being extended or even properly maintained. In contrast, the railroads in the North expanded tremendously, becoming an increasingly more effective transportation network as the years of combat went on. Without iron, the Southern system began deteriorating quickly; as early as 1862, work crews routinely ripped out rails from smaller lines to replace those in the more valuable arteries. By late 1864, the network teetered on the brink of collapse as fewer and fewer trains clattered over the decrepit tracks. When the shooting ended in April 1865, disintegration of the rails and rolling stock was far advanced. Iron indeed had emerged as a still more indispensable mineral resource in nineteenth century warfare, not only as the material of choice for weapons, but just as importantly, necessary for the effective and efficient transport of troops and supplies by the railroads. And the South never had nearly enough.

Minerals in Military History

Walter Youngquist, a noted economic geologist who studies the relationship of Earth resources to nations and individuals, wrote that

> Minerals move civilizations...," further noting that "History has quite properly recorded the progress of civilization in terms of various ages...designated as the Stone Age, the Copper Age, the Bronze Age, and the Iron Age. The use of minerals has provided the material basis for the development of civilization, and has been a major factor in the rise and fall of communities, empires, and nations from the Stone Age to the present. [281]

Precisely when and how humans first began to use minerals and rocks as tools and weapons is not known. The colors and shapes of naturally available items like flint and obsidian (volcanic glass) would certainly have attracted the interest of early peoples to employ them as decorations, simple implements, and cutting tools. Stone- and flint-tipped spears and arrows are among the oldest known instruments of battle.

Salt, or halite as the actual mineral is called, is an excellent example of a natural, widely available substance with numerous military benefits. Its use in preserving food for warriors and providing essential nutrition for them and their pack animals must have been discovered long before written history. The ability of salt to keep meat edible for extended periods allowed military campaigns to take place farther from the homeland or conquered territory where access to livestock was secure.

In addition, this mineral functioned as the most effective wound disinfectant for millennia. It is said that thousands of French soldiers perished during Napoleon's retreat from Russia because a scarcity of salt prevented proper treatment of their wounds.

Salt became so valuable in the ancient world that in many places it functioned as a form of currency. Roman soldiers sometimes received payment in salt, or *salium* in Latin, from which we get the word salary. We still use the phrase "not worth his salt" to indicate someone who isn't doing the job properly. Salt wars were fought in medieval Europe at a time when no nation would begin hostilities without adequate stocks of salted meat for its soldiers and sailors. In America, the Continental Congress took action to ensure a ready supply of salt in the Revolutionary War. With the outbreak of civil war, armies on both sides required immense quantities of salt for the soldiers, horses, and livestock. The gigantic salt-making facilities at Syracuse, New York, and Saltville, Virginia, provided the lion's share of this mineral for the Union and Confederacy, respectively (Fig. 2.2).

Metals are another mineral product that have been enduringly important in warfare since antiquity. Gold and silver commonly anchored the national wealth used to finance military forces necessary for empire building, whereas baser metals—most notably copper, tin, and iron—made the weapons of conquest. Sometime prior to 15000 BCE, humans started working with metals, first fashioning objects made of copper and gold, probably because these two naturally occur in their native (pure) state and do not have to be laboriously separated from worthless rocks.

Fig. 2.2 Salt evaporating ponds and sheds at Syracuse, New York, around 1905. Massive salt works like these in the 1860s made Syracuse the main Union provider of this strategic mineral. Photograph courtesy of the Library of Congress

A variety of cuprous decorations, charms, implements, and weapons have been found dating from the early Copper Age, a transitional time between the Stone and Bronze Ages. Around 4000 BCE ancient people learned the art of extracting non-native copper from rocks rich in that metal by heating the host rocks and melting out the copper—in short, smelting. About a 1,000 years later, coppersmiths discovered that the addition of certain elements, in particular tin, imparted the new alloy with much greater strength than pure copper. The Bronze Age had begun, and soon Mesopotamian soldiers with bronze swords, helmets, and armor slashed deep swaths through the cultures of the Middle East still using unalloyed copper.

The pace of obtaining metals from host rocks rapidly increased in the early Bronze Age. By 3000 BCE, in addition to copper, tin, and gold, metals including silver, lead, and zinc entered into production. The first use of iron, perhaps taken from meteorites where it occurs in the metallic state, may have occurred about then. This wonderfully hard and strong material, however, remained difficult to smelt from ores owing to the high temperatures required. Rare findings of smelted iron artifacts as old as 2500 BCE are known, but widespread use of this metal did not take place until the 1200–1000 BCE time period. The first iron objects proved to be actually weaker and more brittle than bronze, but before long advances in iron metallurgy overcame these problems and made ferrous weapons the choice of Middle Eastern armies.

The rise of classical Hellenic civilization is closely linked with the mineral riches of the eastern Mediterranean region. The nascent Greek city states readily obtained copper from Cyprus and Asia Minor, together with tin, probably from Anatolia (eastern Turkey). Iron deposits, common in the Aegean area, were being worked more intensively during the first millennium BCE as smiths learned the secrets of smelting out the metal more effectively, thus purifying it and making it stronger. Improved technology drove better weapons innovation so that by the seventh century BCE, "...a Corinthian soldier had a bronze helmet, breast-plate, and greaves, a large elliptical bronze shield, a heavy nine-foot thrusting spear with an iron head, and an iron sword" [155]. Thus equipped, warriors from Athens, Sparta, and the other city states of the classical period achieved great martial power, eventually defeating even the mighty Persian hosts bent on ravaging the Greek states.

A century and a half after the Athenian victory at Salamis in 480 BCE, Phillip II of Macedonia gained control of the gold and silver mines in Thrace and used this wealth to subjugate all of Greece. The income from these minerals passed on to his son Alexander, helping him to forge an expanded empire that included the lands of the Persian foes he had at last conquered. Later on the Greek preeminence declined, replaced by the next Mediterranean superpower—Rome. Like Athens before it, Rome depended substantially on precious metals to finance its war-making. Foremost among these were the lucrative silver ores of Spain. In the Second Punic War (218–201 BCE), Hannibal, the Carthaginian commander, controlled those mines for nearly two decades. The seriousness of the loss of income to Rome partly explains why Scipio, the Roman general, attacked Spain despite Hannibal's rampage through the Italian homeland at the same time. Julius Caesar himself invaded Iberia for silver to finance his campaign against Pompey. The Roman commander successfully wrested the Spanish bullion from Pompey's grasp and Caesar's most dangerous rival lost his chance to rule Rome.

The fall of the Roman Empire in the late fifth century CE brought hard times in the West and most of the population was reduced to subsistence farming. It wasn't until the discovery of prodigious deposits of silver, lead, copper, and zinc at Rammelsberg in southeastern Germany in 938 CE that an economic revitalization arose on the continent as the early Middle Ages drew to a close. Centuries later, Spain and Portugal used gold and silver from the New World to fuel their rise to supremacy.

In the 1700s, Great Britain prevailed as the predominant maritime power and used that naval dominance along with her abundant reserves of iron and coal—the two most valuable natural resources in the Industrial Revolution—to become the new economic and military imperium. The British Isles also possessed significant sources of zinc, lead, tin, and copper that had long been known since Roman times. From 1700 to 1850, British lead mines provided more than half of the world's supply. From 1820 to 1840, that nation produced 45 per cent of the global copper, and from 1850 to 1890 from one-third to one-half of the planet's iron. But by the late nineteenth century a new mineral powerhouse was ascending, possessing far more extensive and higher quality assets from the earth.

Rise of the American Mining and Minerals Industry

The story of the American minerals industry originates in Virginia with the first permanent English settlers at Jamestown in 1607. King James I of England sent these hardy pioneers across the Atlantic with the First Charter of Virginia assigning them "Lands, Woods, Soil, Grounds, Havens, Ports, Rivers, Mines, Minerals, Marshes, Fishings, Commodities, and Hereditaments" (heritable property) along the coast of North America. In addition to bringing Christian religion to the native peoples, they had instructions "…to dig, mine, and search for all Manner of Mines of Gold, Silver, and Copper…" and "…to HAVE and enjoy the Gold, Silver, and Copper…" [243]. Of course, such riches must be shared with the Crown, but the message was clear: look for and tap into any and all mineral resources in the embryonic colony.

The newly minted Americans wasted no time fulfilling their charge, erecting a salt works right away and spending time in the chill of an early winter collecting shiny gold-colored grains from the James River sands. Next spring, a barrel of those metallic fragments was shipped out to England, where assayers found them to be pyrite, the mineral called fool's gold that has often been confused with the true precious metal. (In the 1570s, English seafarer Martin Frobisher had made the same mistake three times when he sailed to the New Jersey area and took back tons of pyrite-bearing rock.) Undeterred by their misadventure with pyrite, Jamestown entrepreneurs turned to production of another metal in 1609 when they mined the first iron ore, a marsh and swamp sediment called bog iron. Thirty-five tons of this ore went to England where smelting brought forth iron of "superior quality." In 1619, the initial American iron smelting works commenced operations near present-day Richmond; however, an Indian raid destroyed those facilities three years later.

Virginia's iron industry languished for the next century, while the other British settlements started their own commercial iron-mining and refining endeavors, most notably the Massachusetts Bay Colony in 1646. Throughout colonial times, iron remained the chief mineral product made by the growing coastal enclaves, despite self-interested English efforts to hold down the evolving competition. Iron ore concentrations were fairly common in the original 13 colonies, but most of them were relatively low grade. At the same time, the seemingly endless forests offered stores of cheap wood for charcoal to fuel the smelting furnaces as demand for ferrous tools, implements, and weapons soared in the burgeoning communities, especially along the frontier. New furnaces and forges appeared at a fast pace and the technology of manufacture improved accordingly. Colonists in Pennsylvania made steel in 1732, turned out cannon balls in 1737, and cast cannon for General George Washington's Continental Army in 1776. Iron masters used ore from Lancaster to fashion the long-barreled Lancaster rifle, also known as the Pennsylvania or Kentucky rifle. By 1775, an amazing one-seventh of the world's raw iron, roughly about 30,000 tons, poured forth from American furnaces and forges.

Although iron-making dominated the eighteenth century American minerals industry, explorers located and exploited a variety of other metal troves. Several colonies produced copper, including Pennsylvania, New Jersey, Massachusetts, and Connecticut. Connecticut mines also generated silver, lead, and tin. Lead diggings opened in southwestern Virginia in the 1760s and provided the metal for American soldiers in the Revolutionary War and War of 1812. Prospectors reported gold occurrences in Virginia in the late 1700s, and mining began in the 1820s. North Carolina miners discovered gold in 1793, and that state yielded nearly all of America's gold until 1828.

In that year, the nation's first gold rush got underway when a major find occurred at Dahlonega in the mountains of northwestern Georgia. Georgia ruled gold output in the United States until the California strikes in 1849 almost completely destroyed the eastern gold industry. Dahlonega endured long enough to furnish gold for the Confederacy, but the precious metals available to the Union were much greater, "… thanks to the California mines and to lodes developed in other parts of the West, notably in Colorado and Nevada" [111]. This mineral wealth played no small part in the North's ability to finance the conflict, something their opponents' economy could never do adequately.

The early American mining enterprises involved not only metals but also non-metallic minerals as well. The settlers had to have salt, and at first they obtained it by the ancient method of evaporating seawater at various sites along the coast. As the pioneers moved inland, salt obtained from natural brines (created from ground water dissolving buried salt layers) sprouted up in a number of places, including Syracuse and Saltville. By the time of the Civil War, salt production from boiling underground brines had occurred in eight other states besides New York and Virginia.

The nation's coal industry began when commercial-scale mining of these combustible rocks occurred near Richmond, Virginia, in the mid-1700s. In a short time, cargoes of Richmond coal arrived routinely in Northern ports such as New York, Philadelphia, and Boston, but that dropped off sharply when competition from the Pennsylvania anthracite mines surged in the 1800s. Niter, more typically called saltpeter, for gunpowder manufacture was extremely valuable to the colonists from the earliest days of settlement and more so during the Revolutionary War and the War of 1812. Caves in western Virginia, which included the present state of West Virginia until 1863, provided saltpeter for the nation in these times and into the antebellum years. When civil war broke out, Virginia hosted numerous caverns known to contain niter; established lead, salt, iron, and coal operations were there as well. The fact that Virginia possessed extensive mineral wealth and facilities for its production had not gone unnoticed by Confederate war planners concerned with the resource base needed to fight. Blood and treasure would be spent to protect those irreplaceable mineral industries from harm.

In the first half of the nineteenth century, the mineral operations in the United States continued to expand, albeit much more so in the North than the South. Various factors contributed to this, the most salient being the geological good fortune of the Northern states. Vast deposits of coal and iron had been discovered there in the late eighteenth century, primarily in central and western Pennsylvania. These resources, which included the sizeable anthracite fields in the northeastern part of the state, pushed Pennsylvania to the forefront of American industrial might. Anthracite is a "hard" coal, much higher in energy content than the softer bituminous coals being mined in other regions. The abundance of anthracite—at one point Pennsylvania had 75 per cent of the world's reserves—made possible the higher temperatures needed for "hot" blast iron making in the Northeast, as opposed to the older and less efficient charcoal-based "cold" blast technology dominant in the Southern states. (These two technologies are discussed at length in Chap. 12.)

The growth of a modern coal, iron, and steel industry in Pennsylvania forced massive changes in the North as the 1800s wore on. Manufacturing and heavy industry became concentrated in that part of the nation. An enormous supporting transportation network of roads, canals, and ultimately railroads sprang up. By the advent of the Civil War, five anthracite railroads centered in Pennsylvania had come on line in the Union states. Eventually these hard coal railways brought about the enlargement of harbor operations at ports like Philadelphia and New York. In short, the Industrial Revolution in America during the first half of the nineteenth century progressed at astonishing speed in the Northern states, driven by the high-quality coal and iron deposits located there. These trends would have devastating consequences for the predominantly agrarian South when open warfare with the North burst out in 1861.

Minerals and the Confederate War Effort

On the eve of secession, the Southern states stood unprepared in many ways to bear arms against the rising manufacturing colossus to the north. One basic aspect of the emerging industrialized warfare was creation of a mineral resources base to supply the factories turning the raw materials into the tools and weapons of conflict. Most of the country's pre-war mines, forges, and foundries operated in the free states or territories north of the Mason-Dixon Line. Within the slave-holding section, Virginia had been the foremost mineral producer in the antebellum years and that persisted into the Civil War. Many of these resources resided in the southwestern part of the state which had prolific deposits of lead, salt, iron, niter, and coal. From time to time, the region's mines brought forth some copper, zinc, gold, and silver as well. Michael Lynch, in his study of the Confederate Niter and Mining Bureau, emphasized the importance of the locale: "...southwestern Virginia contained most of the Confederacy's accessible mineral wealth. Other areas in the South had mineral deposits, but none so rich and varied as southwest Virginia" [126].

Although generally not as well established as those in Virginia, mineral industries in some other Southern states did contribute during the war. Alabama coal and ferrous ores aided substantially by feeding the iron-making facilities at Selma as well as mills in Georgia and Tennessee. Gold mines in the Carolinas, Georgia, and Alabama helped fund the martial effort. A copper operation at Ducktown, Tennessee, ended up being the Confederacy's leading source of that metal, necessary for the making of brass and bronze items such as cannon and percussion caps for rifles. The war did not require copper in huge amounts, however, and supplies were usually adequate, even after the loss of Ducktown in late 1863. As war clouds gathered, then, the secessionist states looked most longingly at Virginia to bring her stores of natural resources and advanced manufacturing base to the battle for independence. This she did when she joined the Rebellion on April 17, 1861.

In 1861 nearly all of the Confederacy's heavy industry lay in Virginia in the vicinity of Richmond. A steady flow of raw pig iron smelted from the ores in western Virginia supplied the capital city's complex of forges and foundries. Foremost among these businesses was the gargantuan Tredegar Works, "...the largest industrial base in the South at the beginning of the Civil War ..." and "...the only facility capable of producing major ordnance, iron plate, and iron products in 1861. During the war, other ironworks were developed in the lower South, but Tredegar remained the leading ordnance producer..." [205]. Most of the Southern locomotives, cars, and rails came from Tredegar's technologically advanced shops, as did half of the cannons and perhaps as much as 90 per cent of the artillery shot and shell turned out as the struggle went on.

Tredegar and the other big iron and steel mills burned coal; therefore, a steady supply of that fuel had to be maintained. A few miles west of Richmond were the Midlothian fields, the single large-scale coal mining location in the South. Due to difficult economic conditions, these historic mines had almost completely stopped operating when the hostilities erupted. Soon thereafter, however, wartime demand for coal drove output sharply higher, thereby providing the energy to fire the factories of the Virginia iron and steel industries. Miners extracted and sent east a lesser, but nonetheless valuable, amount of coal from the southwestern part of the state where semi-anthracite deposits had been worked since the late eighteenth century.

Of the mineral resources essential to fight in the 1860s, "Lead was among the most vital... Without it, weapons had no ammunition" [192]. Virginia's lead mines and smelting furnaces at Austinville near Wytheville in the mountains of the southwestern part of the state had been up and running since the mid-1700s. Worked around the clock seven days a week, these large installations constituted the only significant domestic supply of lead in the Southern states. From time to time, the Confederate government ran a handful of much smaller works scattered across the South, but always as backups to the Virginia mines. By the conflict's conclusion, at least one-third of all the lead used by the Confederacy to cast bullets had come from Austinville.

In the next county over from the lead mines sat another critical wartime mineral source—the salt works at Saltville. For armies of the period "… in the days before artificial refrigeration salt was a military necessity of the first importance, since meat could not be preserved without it" [111]. Furthermore, it was also an absolute requirement in the daily diets of the soldiers and animals and for leather fabrication. The Confederacy had a number of salt-making sites when the war came, yet most of these swiftly fell as the North captured the marginal territories harboring them. Saltville grew to be far and away the principal contributor, in the end turning out two-thirds of the South's salt.

Another fundamental component of warfare in the 1860s was nitrate for gunpowder. Nitrate could be refined from a natural organic material called niter, or saltpeter. Saltpeter occurs in a variety of places in the United States, but is most common in the limestone caves of the Southern states. Such caverns are rife in western Virginia and helped provide more niter to the Rebel cause than any other state. When the opening shots thundered at Fort Sumter, the South found itself with very little gunpowder on hand. It became immediately apparent that the Confederacy had to create a reliable domestic supply of niter for the powder mills being constructed. Within a year, an agency to manage saltpeter production expanded into a minerals procurement bureau that did much to keep alive the South's ability to wage war.

In April 1862 the Confederate Congress enacted legislation to establish a Niter Corps within the Ordnance Department. At first, the corps functioned solely to obtain saltpeter to feed the rapidly expanding gunpowder industry. A year later the Congress made the Niter Corps an independent agency and renamed it the Niter and Mining Bureau. This action enlarged the bureau's staff and handed it the responsibility for acquiring not just nitrates but also iron, copper, lead, coal, and zinc. Legislation passed in June 1864 added more personnel, including a maximum of six chemists and six professional assistants, many of them geologists in today's terminology. Among the employees of the Bureau were brothers John and Joseph LeConte, and Nathaniel Pratt, all turning out geological maps as part of their work. Mount LeConte in the Great Smoky Mountains of Tennessee, one of the highest peaks in eastern North America, recognizes the pioneering work of Joseph (and, some think, John as well) on the geology of the southern Appalachian Mountains.

In spite of the Niter and Mining Bureau's best efforts, the domestic minerals industry never reached its full potential, but not because the raw materials weren't there—they were, and in great quantities. What was lacking were sufficient workers to mine and smelt the ores, an adequate road and rail system to carry the output to the factories and distribution centers, and enough soldiers to defend the mineral-producing lands against incursions by the Union. From the onset of hostilities, southwestern Virginia drew special attention from Federal strategists cognizant of the contributions of its mineral products and the railroad that transported them to the Confederacy's capacity to make war. That rail line and the mineral works became the focus of continuing Union assaults throughout the conflagration.

Notes

The first paragraphs of this chapter's introductory section are taken from the general discussion of Virginia's mineral contributions to the Confederacy in Boyle [22] and the emphasis on material resources in industrialized warfare in Kennedy [108]. The growing importance of railroads in the Civil War is based on Hawley [78]. The iron shortages for the railroads and other Southern uses are from Black [18], Ketchum and Catton [112], and McPherson [147]. In the "Minerals in Military History" part of the chapter, the discussion of ancient times through the late nineteenth century draws from Jensen and Bateman [95], Mirsky and Bland [156], Youngquist [282], Mirsky [154], Cowen [35], Craig et al. [36], and Kurlansky [115]. The "Rise of the American Mining and Minerals Industry" opens with material from the Virginia Charter of 1606 which I found online at Yale University as cited in the Bibliography. The rest of this discourse is mainly from Ketchum and Catton [112], Dietrich [45], St. Clair [219], and Mirsky [153]. Also in this section, Hoyle [87] provided an excellent account of the contrast between the Northern and Southern iron and coal industries development. The concluding analysis of "Minerals and the Confederate War Effort" is intended to be an overview and preview of the chapters that follow concerning each of the wartime strategic mineral resources. (In these following chapters, numerous references concerning the individual minerals are presented.) For the present chapter's brief look at the Southern mining and minerals operations, I referred to Boyle [22], Hibbard [81], and Lynch [127] for the broad perspective; the latter is especially good for a treatment of the Niter and Mining Bureau. Summaries of each of the specific minerals are found in a series of articles in the Encyclopedia of the Confederacy by Schroeder-Lein [203, 204] (niter), Robertson [192] (lead), Holmes [85] (salt), Schult [206] (iron), and Hibbard [81] (coal).

Chapter 3
Terrain and a Tale of Two Nations

"To plant an English nation in America"

Jedidiah Hotchkiss, a 19-year-old native of New York state, arrived in Staunton, Virginia, near the southern end of the Shenandoah Valley, in 1847. There he married and settled into life in his adopted state, teaching and continuing his own education by studying geology, surveying, and engineering. His remarkable mapping skills soon emerged when he began to draw topographic maps depicting the local country-side over which he roamed. Once the war commenced, he cast his lot with the Confederacy, and in 1862 received appointment to the staff of General Thomas J. Jackson (after the First Battle of Manassas, known as "Stonewall") as Chief Topographical Engineer. The general immediately called Hotchkiss to his quarters and ordered him to "Make me a map of the Valley...showing all the points of offence and defence" [194]. Jackson found his new chart maker's creations indispensable to his operations, particularly during the Shenandoah Valley Campaign that stamped him as a truly gifted military commander (Fig. 3.1).

After Stonewall died at Chancellorsville in 1863, other Confederate officers profited from Hotchkiss' cartographic talents. He often worked for Robert E. Lee and marched north with him during the Gettysburg campaign. His beautifully drafted plots of campaign routes and battlefields were unsurpassed by any other map-maker on either side for their clarity, comprehensiveness, and accuracy. In the post-war years, Major Hotchkiss helped rebuild Virginia's shattered economy by developing and promoting the state's mineral resources, especially the coal fields in the southwestern region (Fig. 3.2).

The trek of young Jed Hotchkiss took place in a long, wide lowland called the Great Valley, a limestone-floored feature etched out of the middle of the Appalachian Mountains by slow processes of erosion acting over long periods of geologic time. This natural highway within the Appalachians stretches from Alabama to New York and has been traveled by animals and humans for thousands of years. Massive herds of beasts like the eastern buffalo first migrated along this track, then came the Native

© Springer International Publishing Switzerland 2015
R.C. Whisonant, *Arming the Confederacy*, DOI 10.1007/978-3-319-14508-2_3

Fig. 3.1 Portion of the Port Republic, Virginia, battlefield map drafted by Jedidiah Hotchkiss. Hotchkiss prepared the map for General Thomas J. ("Stonewall") Jackson in his 1862 Shenandoah Valley Campaign. Image courtesy of the Library of Congress

Fig. 3.2 Mapmaker Jed Hotchkiss around 1890. Photograph courtesy of the Library of Congress. This talented cartographer served on the staffs of Generals Robert E. Lee, Stonewall Jackson, and Jubal Early during the Civil War

Americans hunting the big game. White settlers in the Great Valley in the eighteenth century translated the Indian name (Athawominee) for their traditional route as the Great Warriors Path. Before long, the old Indian trail had become an important colonial thoroughfare for pioneers journeying within the mountain ranges. In time, roads followed the Valley floor from Pennsylvania through western Maryland and the Shenandoah Valley of Virginia, and on into the deeper southern mountains. Passing through southwestern Virginia and into northeastern Tennessee, the route connected with the Wilderness Road that carried settlers west through Cumberland Gap into the Kentucky frontier and the unsettled territories beyond.

No other colonial pathway through the Appalachian back country was more important to the development of the nation than the Great Valley corridor. Beginning in the 1740s and continuing well into the next century, large numbers of Scots-Irish (Jed Hotchkiss' family was of Scottish descent) and Germans, typically from Pennsylvania or New York, moved south along the Great Valley into western Virginia. Many made their homes in the fruitful Shenandoah bottomland, whereas others went farther on the Valley road into the southwestern part of the state. There they established farms and, by the late 1700s, founded industries that included extraction of the lead, salt, iron, and saltpeter resources. Settlements and towns sprang up in Southwest Virginia, growing prosperous from the trade and commerce moving along the busy frontier passage. In time, the Virginia and Tennessee Railroad came down the Valley, adding a critical new element to the regional transportation system. By the 1860s the railway had become an indispensable link between the agricultural and mineral products of the Valley and the factories and population centers of eastern Virginia.

Geologic History and the Formation of Topographic Provinces

The making of the divergent landforms of eastern America extends back into the deep geologic past. Early in the Paleozoic Era, about 540 to 450 million years ago and long before the Appalachian Mountains elevated, a thick sequence of sediments piled up along the eastern (present-day coordinates) edge of the ancient North American continent known to geologists as Laurentia. Some of these layers were deposits of beach and river sands, and others limey accumulations similar to those presently carpeting tropical seafloors in places like the Bahama Islands. In a process persisting to this day, Laurentia and the other ancient continents wandered about, colliding with smaller land masses such as islands and volcanic arcs or merging with each other from time to time.

Two such impacts occurred in early and middle Paleozoic time, each one welding new landmasses we now call the Piedmont and Blue Ridge provinces onto the eastern margin of Laurentia. These encounters put intense pressure and heat on the rocks and sediments in the crushed zones that compacted the loose sediments into

solidified rocks such as sandstone and limestone. Many of the rocks caught in the
core of the deformed belts were transformed into metamorphic rocks like slate,
schist, and gneiss; others finally melted and then cooled to form igneous rocks such
as granite or rhyolite.

Late in the Paleozoic, about 250 million years ago, convergence of the drifting
continents resulted in the assembly of the supercontinent Pangea. During the stitch-
ing together of Pangea, a continent-continent collision between ancient North
America and Africa compressed and uplifted the rocks between them to build the
Appalachian Mountains. Originally, the freshly minted mountains would have had
alpine heights, but erosion since then has reduced the ranges to the much lower elevations seen today. Approximately 200 million years ago, another cycle of supercontinent breakup and assembly started when Pangea rifted apart. Streams and rivers
cascading from the majestic Appalachian peaks stripped off torrents of gravel, sand,
and mud, carrying them east to pile up along the coast of the recently born Atlantic
Ocean. Since that time, a thick stack of this sediment has constructed an extensive
coastal plain south of the New England states along the Atlantic and Gulf margins.

The mountains and lowlands of the Mid-Atlantic and Southern regions brought
forth by this geologic history are clustered into five topographic (or physiographic)
and geologic provinces—the Coastal Plain, Piedmont, Blue Ridge, Valley and
Ridge, and Appalachian Plateaus from east to west (Fig. 3.3). In Virginia, the
Coastal Plain is referred to as the Tidewater country because the daily tidal changes
along the coast affect river levels far inland. The geologically young unconsolidated
sediments still washing off the Appalachian Mountains make up this province. With
an absence of resistant rocks to create hills or ridges, the ground surface eroded onto
the loose detritus displays few differences between the highest and lowest points.

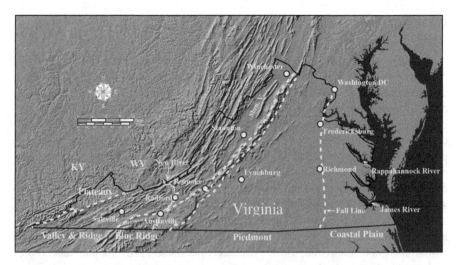

Fig. 3.3 Map of the geologic and physiographic provinces of Virginia. Note the Fall Line
separating the Coastal Plain and Piedmont, and the Shenandoah Valley ("Shen V.") in the northwestern section

Water does not drain readily off this nearly flat terrain, therefore bogs, swamps, and rivers prone to frequent flooding abound.

The Coastal Plain extends west from the Atlantic shoreline to its boundary with the Piedmont, a junction known as the Fall Line. The Fall Line is a zone of rapids and falls created in the rivers as they flow from the harder, more resistant rocks of the Piedmont into the softer Coastal Plain sediments. The Piedmont surface is an expansive, gently undulating platform that gradually rises higher from the Coastal Plain toward the Blue Ridge. Many rivers, the Rappahannock and James being two of the largest, flow down that slope and across the Coastal Plain, ultimately debouching into the Atlantic. The low hills and shallow valleys of the Piedmont lie atop the igneous and metamorphic rocks formed in the mountain-building events of long ago. Though these bedrock types are hard, 200 million years of erosion have worn them into low, stumpy remnants of their former grand heights.

The Fall Line contact between Piedmont and Tidewater is a geologic phenomenon that had enormous consequences for the settlement of eastern America. Streams coursing across the Piedmont have shallow, rock-strewn channels due to the difficulty of cutting down into the ancient stone basement. These same rivers, however, become deeper and wider when the more erodible Tidewater sediments are encountered. This zone of transition and turbulence in the rivers where the Piedmont meets the Coastal Plain is the Fall Line.

The numerous rapids and waterfalls caused by the rivers suddenly dropping to a lower elevation along the Fall Line presented a threatening barrier to riverboat navigation farther upstream. At such a daunting point, many pioneers chose to stop and build their homes, shops, and farms. Though an impediment to travel, the dangerous plunging waters yielded abundant energy that the enterprising arrivals found extremely useful to power their mechanical devices and the little communities continued to grow. Several of the early settlements and trading posts born along the Fall Line later turned into major eastern cities, including Philadelphia, Baltimore, Washington, and Richmond.

Abutting the western edge of the Piedmont, the Blue Ridge province is a relatively long and narrow mountain range made of older, harder rocks like the Piedmont, but with much higher elevations due perhaps to more recent uplift. Here and there water gaps and dry passes slice through the Blue Ridge and connect the Piedmont with the Great Valley, rendering movement through the mountains much easier. Although graced with splendid scenic vistas, the rugged topography in the Blue Ridge made eking out a living there a challenge that prevented large scale settlement from developing.

West of the Blue Ridge lies The Valley and Ridge, characterized by steep-sloped, long ridges lying between deep valleys aligned generally northeast-southwest. This portion of the mountain system is made of a variety of sedimentary layers deposited in the days of ancient Laurentia when shallow seas often covered its margins. Two of the most common rocks are sandstone and limestone, products of the old coastal sands and tropical limey deposits. Folding and faulting during mountain building heaved a broken array of these varying kinds of strata to the surface where they wear away at different rates. Limestone erodes fairly rapidly into lowlands, especially in a wet climate like that of the eastern United States. In contrast, sandstone, being more resistant and losing surface expression at a slower pace, forms hills. This succession of parallel alternating limestone-floored valleys and sandstone-crested ridges, looking like gigantic rock waves frozen in a sea of stone, extends for most of the length of the Appalachian chain. The Great Valley, located in the easternmost part of the Valley and Ridge, is by far the biggest of these limestone depressions.

The Great Valley nestles between the lofty granite-ribbed peaks of the Blue Ridge and the tall sandstone heights of the Alleghenies to the west. In western Virginia, the widest part is the northern extent known as the Shenandoah Valley, named for its celebrated river that drains into the Potomac. The Great Valley exists because of an unusually broad and thick package of limestone beds lying beneath it. Calcium carbonate, the same material that composes clam shells, makes up the limey deposits and dissolves in the slightly acidic ground waters continually seeping through the subsurface. In a sense, the limestone "melts" away, leaving a trough several hundred feet lower than the hard rock ridges on either side. Constant dissolving of the limestone riddles it with thousands upon thousands of underground caverns, Endless and Luray being two of the more famous in western Virginia. Saltpeter, a natural nitrate-rich deposit that can be refined into the main ingredient in gunpowder, occurs in many of these caves.

The smallest of the Virginia physiographic divisions is the Appalachian Plateaus found in the southwestern counties. This province is comprised of sedimentary rocks similar in age to those of the Valley and Ridge, but here the rock layers are nearly flat lying instead of tilted at high angles. The table top orientation of the strata presents a uniform surface of erosional resistance to running water, allowing the local streams to flow in any direction as long as it is downhill. The result is a multi-branching, tree-like network of flow finally merging into trunk rivers exiting down the regional slope. This virtually random drainage pattern creates a landscape intricately dissected into numerous isolated steep-walled canyons with narrow valley floors separated by level ridge tops. Put another way, this is the classic "hills and hollers" country of Appalachia, typical of Southwest Virginia and adjacent West Virginia and eastern Kentucky.

The terrain fostered by the geologic processes that built Virginia and the southeastern United States loomed large in Civil War military actions. The Union and Confederate capitals were separated by only about 100 miles of gently rolling, well-drained Piedmont topography, good fighting ground where the armies of Blue and Gray could be readily maneuvered and supplied. This area would see almost constant combat as the contending forces shifted back and forth. The fertile level floor of the Shenandoah Valley, breadbasket of the Confederacy and an excellent pathway for military operations by both sides, turned into another primary battlefront as the conflict went on. In southwestern Virginia, the Great Valley, besides being the principal travel route through the locale, was the site of nearly all the mineral works and the Virginia and Tennessee Railroad. Here, the battles for control of the steel rails and the raw materials manufacturing sites raged on throughout the war.

Plate Tectonics and the Discovery of America

The profound impact of the physical geography and geology of eastern North America on its exploration and development started with the very discovery of the continent by Caucasian sailors. That discovery can be attributed to its geographic location opposite those nations of Western Europe with the most advanced seagoing abilities. Once mariners in Portugal, Spain, France, and England began to venture farther and farther west from their shores to find the Orient, discovery of an unexpected world only a few thousand miles across the ocean became inevitable.

According to the principles of plate tectonics, the North Atlantic's present width of 3,000 miles, relatively narrow for a major ocean basin, is due to the creeping pace of seafloor spreading that brought it into being. Seafloor spreading is the fundamental geologic process in the formation of ocean floor crust. In this mechanism, molten basalt rises from the planet's interior, breaks through continents, and floods out along volcanic mid-ocean ridges to create brand new seafloor between the landmasses. The continental fragments then move away from each other, making room for the widening or "spreading" ocean basin. Seafloor spreading in the North Atlantic currently is driving apart the European and American coast lines at less than an inch per year, an exceptionally sluggish speed when compared to spreading rates as high as six inches annually in other ocean basins. Although the North Atlantic cracked open about 200 million years ago, the slow building of its sea bed crust has not had enough time to generate a broad ocean basin that might have disheartened and turned back those first European explorers voyaging west.

Until the fifteenth century Europe depended on overland trade routes from the Orient to bring in spices, silk, jewels, tea, porcelain, and other valuable commodities. In addition to these wares, innovative technologies such as paper and printing, the compass, and gunpowder traveled to the West across the Eurasian land mass. That all changed abruptly in 1453 when Muslim capture of Constantinople shut off the last land link to the Far East. Prices for Oriental goods shot up and another connection to the East had to be found. With combined hope and desperation, Europe turned to the seas for different ways to southeastern Asia. In 1492, Columbus at last crossed the Atlantic, blown by favorable easterly trade winds and using the Canary and Azore Islands as stepping stones. Yet instead of finding the Orient, he splashed ashore onto an entirely new world. In fact, earlier Norse explorers had reached North America long before Columbus, but no real interest in the cold, harsh lands found by them had followed. Now, at the dawn of the sixteenth century, England and France entered the ferocious competition for the resources of the rediscovered continent and its bordering waters.

The distance from the British Isles across the Atlantic to eastern Canada is little more than half the length of Columbus's voyage, but this northern route has much fiercer weather and goes against the prevailing westerly winds. Despite the more difficult sailing, by the early sixteenth century Portuguese and French ships were exploiting the bountiful fishing grounds that came to be known as the Grand Banks of Newfoundland. Seventy years later, an estimated 350 fishing vessels from those nations, as well as England and Spain, routinely hauled in immense catches of cod from the shoals off eastern North America. With the rise of lucrative markets for not only America's fish but also its furs and timber, settlement was bound to follow. In the late seventeenth century, Sir Walter Raleigh spoke for many with his notion "...to plant an English nation in America" [208]. That desire came to permanent fruition at Jamestown in 1607. Within a century British settlements dotted the coastal strip from Nova Scotia to Georgia.

Rocks, Soil, Topography, and the Growth of Sectionalism

The European immigrants who colonized the edge of the new northern lands did not stay there. The many streams coursing into the Atlantic and its largest embayment, the Gulf of Mexico, offered natural avenues for interior exploration and settlement. The big, lengthy rivers found south of New England have lower reaches that tend to be deep and wide where even boats that sailed across the ocean could travel far upstream. Furthermore, the numerous streams in the central and southern Atlantic coastal zones flow across a low-lying plain several hundred miles across in the southernmost stretches. Dipping gently down to the ocean, this expansive platform narrows to the north and disappears beneath the sea off the northeastern coast. The early navigators went up the inviting Atlantic coast riverine highways and into the continental interior seeking minerals, furs, timber, and other resources that might bring wealth to them and the mother country. Shortly thereafter, settlers arrived, looking for virgin lands to homestead.

Basic differences in geology and landforms between the Northern and Southern regions of the North American Atlantic coast held tremendous ramifications for the future nation only now beginning, particularly as concerned its developing economies. In New England, recently vanished Pleistocene Ice Age glaciers had cut down to the bedrock, leaving a thin rocky soil in their stead. This bouldery mantle required much effort to remove the huge stones prior to cultivation. Once farming commenced, the natural nutrients quickly gave out using the agricultural methods of the times. The poor soils and absence of a coastal belt conducive to capacious commercial farms early on steered New Englanders away from agriculture and toward the ocean fisheries and inland fur trade for their livelihoods.

Conditions in the Mid-Atlantic and Southeast, however, were very different. Here the mild climate, vast areas covered by thick, productive river-laid sediments, and many streams to transport the produce made farming the most attractive wealth-generating enterprise. Expansion of agriculture across the extensive lowlands was easy, and spacious estates grew rapidly. By 1685 Virginia alone had lands under cultivation with a total geographic extent equal to all of England but worked by a mere fraction of the English population.

The ocean and rivers were the earliest large physical features affecting the settlement of the Atlantic seaboard colonies, but soon another great geographic and geologic determinant became important—the Appalachian Mountains. This ancient mountain system, among the world's oldest that still have topographic expression (those formed earlier have been worn down to their roots and no longer show any elevation above the surrounding countryside), extends some 1,500 miles from central Alabama to Maine and beyond. For most of that length, east-facing steep slopes abruptly rise several thousand feet above the Piedmont foothills butting up against them. This imposing wall of rock offers few passages through it, severely deterring further migration into the interior. Thus bounded by sea and mountains, the fledgling "English nation in America" took root and flourished.

The isolation created by the ocean and highlands wrought pronounced and enduring effects on the people of the nascent states. The Atlantic and Appalachians provided security from outside threats and fostered interaction and bonding among the colonists. As the population increased, common political and social attitudes and a corresponding sense of unity appeared. The difficulty of penetrating the mountain barrier had another notable impact: removal of the temptation to expand into the almost limitless prairies of the Mid-continent. The French and Spanish had succumbed to this lure when their traders and trappers scattered over the enormous inland wilderness without establishing the network of towns and cities needed to anchor a national presence.

For 150 years the Appalachians held the mostly British settlers hemmed in on the margin of the continent and the Atlantic kept them physically separated from England nearly 3,000 miles distant. As a result, a strong frontier spirit, founded on a growing sense of independence, emerged. With the passage of time, this engendered increasingly frequent clashes with the faraway homeland government over interference in colonial affairs. In 1775, the once loyal Americans could brook no further meddling from abroad and the Revolution broke out.

With freedom won in 1783 and England no longer a common enemy to bind North and South together, differences in economic, social, and political institutions, brought on by sharp distinctions in climate, topography, and soils, intensified. By the 1790s, American geography books were calling attention to the growing sectional divergence: "The northern and southern states differ widely in their customs, climate, produce, and in the general face of the country. The fisheries and commerce are the sinews of the north, tobacco, rice, and indigo of the south; the northern states are commodiously situated for trade and manufacture, the southern to furnish provisions and raw materials" [24]. New England's rocky glacial soil and lack of coastal low country had indeed swayed the industrious folk away from agriculture toward manufacturing, fishing, and trade; consequently, commercial capital investment flowed into those sectors. The firm ties to the sea and the abundance of steep gradient, fast flowing streams to power the mills and factories sprouting up created little desire for westward expansion, an impetus already felt strongly in the South.

In New York and Pennsylvania, although there is more arable land than in New England, a wide, flat lowland marginal to the sea and so inviting for creation of grand estates is nearly absent. Therefore, the economy and culture here more closely resembled that of the most northern states. In contrast, Virginia and the states farther south possessed extensive farming space on the Atlantic and Gulf coastal plains that fostered the establishment of sizeable farming endeavors. At the outset, a distinctive Southern agriculture evolved that "…depended upon adaptation of the slave plantation system to the Indian crop tobacco, and it took place in the early seventeenth century in Virginia and Maryland" [174]. The growing of the "Indian crop" returned immense profits; it was "…the means to wealth in Virginia…" [90] and soon in the Carolinas as well.

But raising tobacco also faced a vexing problem—continuous cultivation of the plant depleted the fertility of the river bottom soils in just a few years. In due course the exhausted fields had to be abandoned and given time to recover their productivity. As a result, the Mid-Atlantic and Southern growers constantly sought fresh territories, which at first led to simply moving upstream along the plentiful rivers coursing over the coastal flats, then spreading out into the low valleys between the major streams. In time, these lands filled up, and longing eyes looked toward the Piedmont, the open ground between the seaboard platforms and the tall peaks farther west.

In 1705 Francis Makemie, an Irish clergyman who had traveled in North Carolina, Maryland, and Virginia establishing Presbyterian churches, called attention to the potential of developing agriculture in the Piedmont. In a published work addressed to the Governor of Virginia, he wrote that "The best, richest, and most healthy part of your Country is yet to be inhabited above the falls of every River, to the Mountains" [121]. In fact, tobacco farming did exceedingly well in that region's subdued topography and deep soil cover, and the big estates dispersed across the area as the eighteenth century wore on. By 1800 most of the tobacco crop grew in the Piedmont or in the adjacent upper coastal plain.

A landmark event that spelled the end of tobacco's dominance on Southern plantations had already occurred, however, in 1794 when Eli Whitney patented his cotton gin. Now much easier to turn into useable fibers, "King Cotton" proved to be even more profitable than tobacco and began its long reign. With nearly all of the arable space in the southeastern United States coming under cultivation, many more slaves were required, rooting that practice more deeply into Southern commerce and culture. Inevitably, planters wanted to expand into untilled lands for profit and to rest the ones they owned, but a formidable topographic barrier had long stood in their way.

Prior to the Revolutionary War, the Appalachian Mountains blocked the westward expansion of large estate agriculture. The few hardy pioneers who dared press on through the infrequent mountain passages set up small frontier subsistence homesteads. Yet beyond the Appalachians lay enormous tracts of relatively flat fertile lands that could be made into farming cornucopias. The British defeat of the French on the Plains of Abraham at Quebec in 1759 settled the question of which nation would dominate the North American continental interior. Twenty-two years later, Washington's victory at Yorktown secured independence for the United States and sovereignty over the Mid-continent prairies. In the southeastern part of the new nation, waves of settlers, especially land-hungry planters, now passed through and around the Appalachian chain to populate and farm the rich alluvial sediments of the Gulf Coast and Mississippi Valley territories. This meant more plantations, more slaves, and a growing number of cultural collisions that would burst into open warfare in 1861.

The pre-war Old South, however, was far from a unified entity of like-minded people sharing the same social views and political goals. Paralleling the growing split between the North and South, two notably different realms of opposed economies and cultures within the Southern states emerged—the so-called Upper and Lower South. The "Upper" and "Lower" designations suggest a zonation based strictly on latitude; more accurately, the distinction is mainly a function of terrain, soil, and climate. The Lower South developed much more completely as "…a land of cotton and slavery, a land dominated economically by the plantation type of agriculture…" whereas the Upper South became "…primarily the domain of the slaveless yeoman farmer, an area largely devoid of cotton and the other subtropical cash crops" [174].

In reality, the economy and societal attitudes of the Lower South existed on the Coastal Plain and Piedmont lowlands extending from eastern Maryland and Virginia down through the Carolinas and into Georgia, Alabama, and (minus the Piedmont) Florida. From there the Lower South swung around the southern end of the Appalachians and extended across the broad prairies of the Mississippi Valley as far west as eastern Texas and Arkansas. The northern end of this section was the southern tip of Illinois, for years known as "Little Dixie."

The Upper South consisted of the mountains of Virginia, North and South Carolina, Alabama, Georgia, Tennessee, and Kentucky. Throughout the Civil War, the areas within the Upper South remained only marginally loyal to the Confederacy and often supported the Union cause not just politically but with men and arms as well. Of great importance was the fact that nearly all of the Southern mineral resources, the very backbone of the Confederate war machine, lay within the contentious mountainous lands.

Effect of Landforms on Virginia's Cultural Development

Virginia's diverse terrain is a microcosm of the Upper and Lower South dichotomy in that all of the topographies so influential in shaping the history of the southeastern United States are present in that one state. Not surprisingly, then, Virginia's pre-Civil War history reflects that of the larger Southeastern section. The arable lowlands along the seacoast experienced the first farming, road building, establishment of towns and cities, and other activities of civilization. Shortly after settling Jamestown, English pioneers left the water's edge, moving to the west across the Tidewater flatlands and into the rolling hills and valleys of the Piedmont over the next two centuries. The Appalachian barrier held back those early Virginians for many years, resulting in a fairly homogeneous culture in the founding region. Politically and socially, the inhabitants of the eastern part of the colony grew to be very similar to those on the vast, slave-holding estates expanding across the Piedmont and coastal low country in the Deep South.

In mountainous western Virginia, a much different story unfolded. As noted earlier, beginning around 1740, significant numbers of people of various national origins, including Dutch, Huguenot French, English, and particularly Scots-Irish and German, migrated south primarily along the Great Valley corridor and stayed to work the virgin soils in the Shenandoah bottomland. Used to more precipitous terrain like they had known back in the old country, many of these sturdy folk ventured into the mountains where a distinctive culture took shape. With little level acreage available, smaller subsistence farms with few or no slaves took hold, and with them political and social outlooks far different from those in Tidewater and Piedmont Virginia. Instead of a highly stratified society dominated by aristocratic wealthy planters, a more democratic and independent-spirited culture evolved. When secession from the United States at last came, the mountain people felt little connection to the firebrands of rebellion on the eastern plains.

Nevertheless, this was not true for all of the highlands area west of the Blue Ridge. The Great Valley, that expanse of relatively flat ground with fecund black soils worked by slaves, strongly resembled the productive domains of the cotton kingdom South. Moreover, this open topography had always served as a prime conduit of travel, attracting early road builders to put their turnpikes here and later on railroad men to lay their tracks along the Valley floor. With much good agricultural land available and the presence of a well-established transportation network, larger farms with increased slave populations proliferated. These were much more akin to the plantations in the Piedmont and Tidewater than to the modest yeoman plots in the surrounding hills. In 1863, when a significant portion of Virginia's mountainous Northwest split off to form the Union state of West Virginia, only two counties, both in the Shenandoah Valley, chose to remain a part of Confederate Virginia. Not a single county in the southwestern Great Valley joined the separatist movement.

Notes

The biographical information on Jed Hotchkiss is taken from Roberts [191], Miller [151], and Robertson [195]. The next section on "Geologic History and the Formation of Topographic Provinces" in the eastern United States is essentially current geologic thinking on this topic merged with my own ideas developed over 30-plus years of studying and teaching about the connections between the region's geology and human history. The basic geologic principles in this chapter—plate tectonics, weathering and erosion, landform development, and geologic history— can be found in any standard geology textbook; two of my favorites are Hamblin and Christiansen [75] and Tarbuck and Lutgens [231]. They are comprehensive and eminently readable; I taught geology courses from them for many years. General references for the geology and topography of Virginia's five provinces are Dietrich [45], Frye [68], and various publications of the Virginia Division of Mineral Resources, most of which are available online as referenced in the Bibliography. The following sections on the "Discovery of America," "Rocks, Soils, Topography and the Growth of Sectionalism," and the "Effect of Landforms on Virginia's Cultural Development" are based on material extracted from Semple [212], supplemented by Brown [25], Paterson [175], Hudson [91], and Linklater [122]. Some comments about Ellen Churchill Semple's work: Her insightful book, *American History and its Geographic Conditions*, was first published in 1903. The version I have is a revision (with Clarence F. Jones) published in 1933. I discovered this book many years ago and it has since profoundly influenced my work in connecting terrain and geology to American history, especially the Civil War. Semple was accused of a simplistic kind of "environmental determinism"—and certainly other factors besides climate, soil, and topography affected our national history—but her meticulously detailed emphasis on the links among latitude, physical features, and the human history of the country is well worth considering. Her writings continue to be read because they so clearly relate the powerful shaping forces of the landscape (and, by extension, geology) to the discovery and settlement of America and the development of its agriculture, industry, political institutions, and culture.

Chapter 4
The Land They Fought For

"A political peninsula projecting into a sea of hostile territory"

Visitors to Gettysburg battlefield almost always grasp immediately what the fighting was about: which side would control the roughly three-mile-long piece of high ground known as Cemetery Ridge (Fig. 4.1). For three fearfully bloody days in July 1863, Union General George G. Meade's Army of the Potomac sat atop this north-south trending spine of rock while Robert E. Lee's Army of Northern Virginia tried and failed to take it from them. Perhaps the single most important piece of contested land in the war, Cemetery Ridge is there thanks to geology.

Gettysburg is located in an enormous geologic feature within the Piedmont called a Mesozoic Basin, so named for the era of time in which it formed. A disconnected string of these basins extends from Georgia to Connecticut, a heritage of the tearing apart of Pangea 200 million years ago. As the continents separated, immense chunks of what would be eastern North America dropped downward into the stretching crust, forming rift valleys much like the ones in East Africa today. The fault troughs filled with river and lake sediments, and dinosaurs left their footprints in the muds along the flood plains and swampy shorelines. Beneath the valley floors, hot melted igneous rock (magma) squeezed into the layers of sediment, from time to time bursting through to the surface and building volcanoes.

It is this diverse assemblage of sedimentary and igneous rocks that gives Gettysburg its distinctive topography. Weathering and erosion have carved out Cemetery Ridge from a thick mass of the magma, now cooled and hardened into a granite-like stone called diabase (Fig. 4.2). The diabase is intruded between beds of shale, a relatively soft rock made of compacted silt and clay sediments. The diabase is far more resistant to erosion than the shale, leaving the exposed ragged edge of the igneous sheet standing higher than the gently undulating landscape formed on the adjoining sedimentary strata. From Cemetery Ridge, the relatively smooth slopes on the shale descend west to a subdued rise underlain by a thinner diabase

© Springer International Publishing Switzerland 2015

R.C. Whisonant, *Arming the Confederacy*, DOI 10.1007/978-3-319-14508-2_4

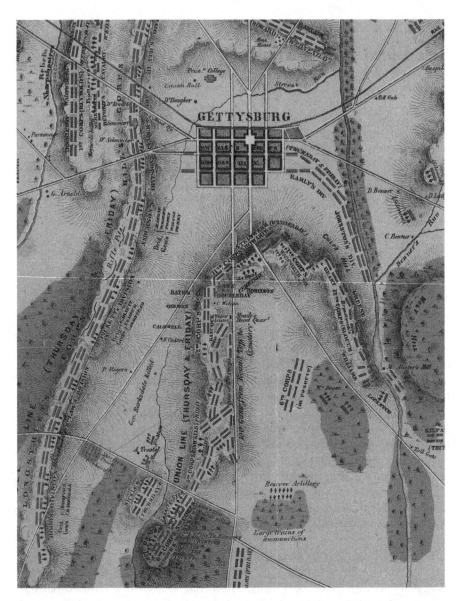

Fig. 4.1 Map of Gettysburg Battlefield. The "fish-hook-shaped" landform south of the town is Cemetery Ridge held by the Union. The long, linear feature running along the left side of the map is Seminary Ridge controlled by the Confederates. Image courtesy of the Library of Congress

Fig. 4.2 Diabase exposed on the crest of Little Round Top near the southern end of Cemetery Ridge. The view is looking west toward Seminary Ridge (lined with trees in the middle distance)

intrusion. This is Seminary Ridge, about a mile distant from the top of Cemetery Ridge. Confederates occupied this low feature, giving them literally an uphill battle to fight for the commanding prominence held by the Union.

The Battle of Gettysburg unfolded into three large and ultimately unsuccessful Rebel attacks. On July 1, Lee assailed Culp's Hill where the United States army had anchored the northern end of its lines; as darkness fell, the onslaught was repulsed. Next day, the men in gray tried to turn the southern end of Cemetery Ridge at Little Round Top, but once more the Union beat back the assault. On July 3, frustrated by the failed flanking tactics, the Confederate commander sent General George Pickett's division forward over the broad shale plain rising toward the center of Cemetery Ridge. With virtually no place to take cover on the bare killing field, the waves of Virginians melted away in the fiery storm of lead and iron pouring from the Federal lines.

When the carnage mercifully ended that evening, 51,000 soldiers had been killed, wounded, or declared missing at Gettysburg. Meade's army still held the heights etched out on the diabase intrusion and the Union was saved. Yet that topographically advantageous location had its own geological drawback that proved costly to the defenders perched on it. Because the igneous rocks are erosionally resistant, the soil is thin and the diabase crops out all along the ridge crest. With hard rock jutting out or just a few inches beneath the ground surface, the Northern troops could not dig in and fortify their lines with protective trenches and earthen walls. The artillery shells bursting around them in the barrage before Pickett's charge and the following infantry rifle fire from the Virginians exacted an awful price for the elevated but exposed position. The Army of the Potomac suffered 23,000 casualties, nearly as great as the Confederate loss of 28,000, and an unusually high number for the defending side in a major battle.

Impact of Terrain on Two Battles in Southwest Virginia

Throughout the history of warfare military commanders have recognized the importance of terrain to the success or failure of their armies in the field. In the words of military geographers Harold Winters et al., "Time and again terrain plays a powerful role in conflict" [272]. The land surface over which soldiers advance, retreat, or fight is always the result of two basic determinants: geology and climate. Harder (more difficult to erode) rocks like granite and sandstone stand as rugged high ground. In contrast, softer (easier to wear away) rocks such as shale or limestone produce broad, fairly flat lower terrain. Climate strongly factors into landscape formation, for example as with limestone, a rock that readily dissolves away to make valleys in the well-watered eastern United States yet supports ridges in the arid western regions.

In the American Civil War, whether on the shale plains at Gettysburg and Manassas or the limestone platforms at Antietam and Chickamauga, fighting men suffered terribly on the open topography produced by the underlying soft rocks. On the other hand, Union troops massed along the diabase at Cemetery Ridge and Confederates atop the sandstone forming Marye's Heights at Fredericksburg had the huge military benefit of nearly invincible positions granted to them by the resistant rocks below their feet.

These same rules of terrain development and its significance for soldiers in the great battles held true for the campaigns and engagements in southwestern Virginia. At the Battle of Cloyds Mountain in May 1864, the Confederate defenders set their infantry and artillery on a low ridge of rock fronted by a creek and its small floodplain. The Union attackers had to wade across the little stream and unobstructed flat floodplain under fire; this they did but at a considerable cost in casualties.

One day later, the Federals advanced to the prime objective of the invasion, the railroad bridge spanning the New River at Central Depot (present-day city of Radford). Arriving on the north side of the river put them on tall rocky bluffs looking down on the bridge and the Confederates arrayed on a broad alluvial plain. In the artillery exchange that ensued, the Northern gunners had the decided advantage of plunging (downhill) fire and seeing where the shots struck. The Confederates, on the other hand, could not observe the Union troop and gun positions beyond the cliff tops and thus could not accurately aim their fire at these targets. Geology added another element as the big guns roared. Colonel Rutherford B. Hayes, one of the Union brigade commanders, ordered his men to take cover in the numerous sinkholes around them. These geologic depressions, common in the limestone and dolomite strata forming the bluffs, gave the only topographic concealment available during the cannon duel.

In fall 1864 a Federal incursion targeted the salt works in Southwest Virginia, leading to the October 2 Battle of Saltville where the terrain presented a challenge to the success of the mission. The Southerners occupied fortifications on the steep slopes and hilltops shielding the valley in which the salt-making facilities lay. In the course of the fighting, the Union attempted to force river crossings at two fords on the North Fork of the Holston River. After repeated assaults could not overcome the topographic barriers, the Northern commander withdrew his columns. The river and the ridges encircling the salt wells and furnaces made an impregnable fortress that day and the South's main salt production center went unharmed. Nevertheless, the Union returned in a few months, this time using gaps through the hills to enter the valley and demolish much of the mineral operations. In the Civil War, as in all wars and battles big and little, terrain was a constant and immutable factor. Wise commanders always gave it careful consideration.

Rocks and War: Corridors and Barriers

The physical features and geology of eastern North America had much to do with the creation of the two separate Northern and Southern sections that came to blows in 1861. Geographer Ellen Semple expressed it thusly: "The question of slavery in the United States was primarily a question of climate and soil, a question of rich alluvial valley and fertile coastal plain, with warm moist, enervating climate versus rough mountain upland and glaciated prairie or coast, with a colder, harsher, but more bracing climate" [209]. In the Civil War, landforms just as profoundly impacted strategy and tactics on all battlefronts. On a grand scale, the Appalachian chain was a fundamental topographic impediment to the passage of armies, effectively splitting the arena of military operations into Eastern and Western Theaters. In the end, the successful Union efforts west of the mountains amounted to a gigantic flanking movement that took the Confederacy in the rear. The fall of Chattanooga in Tennessee to Grant in late 1863 followed by troops under Sherman in 1864 pushing through and around the southern end of the Appalachians and into Georgia irrevocably doomed the Confederate hope of independence.

The Union victory at Vicksburg had earlier cut off the trans-Mississippi states, and Sherman's capture of Atlanta and March to the Sea separated the Gulf states from those of the Atlantic. Chopping up the Confederacy into three essentially isolated pieces enabled Northern armies to more rapidly gain control of each. In the most critical segment—Virginia and the Carolinas—relentless pressure by the Federal forces on Richmond from both northern and southern directions at last drove the Confederates from their capital city and into surrender at Appomattox.

Both sides realized from the start that the decision had to be won in the East. Virginia, the northernmost of the seceding states and the primary Union entry point into the Confederacy, became the cockpit of combat. The National Park Service recognizes 122 battlefields in Virginia, about one-third of the total fought upon in the entire conflict and over three times as many as second-place Tennessee, the principal battlefield state in the West.

The emergence of Virginia as the war's main battleground is due in no small part to its precarious geographic position, described by Semple as "...a political peninsula projecting into a sea of hostile territory" [210]. On the eastern flank, the Union navy controlled the waters of the Chesapeake Bay for the duration of the war following the battle of ironclads USS *Monitor* and CSS *Virginia* in March 1862. With much Federal political and military coercion, Maryland to the north and Kentucky to the west stayed in the Union where they were joined by the new state of West Virginia in 1863. Therefore, half way through the hostilities the North had secured all of the approaches to Virginia save for its southern border. For the next two years, the Union army pounded away in that state, striving to subjugate it and advance deeper into the Atlantic seaboard Confederacy.

Some fighting occurred in all five of Virginia's physiographic provinces, but most of the principal battles took place within the Piedmont and Great Valley. These two low, relatively level areas served as military corridors, places where masses of troops can be more easily handled and supplied. The other three provinces—the Coastal Plain, Blue Ridge, and Appalachian Plateaus—and the mountains of the western Valley and Ridge were barriers, locations where the more difficult terrain severely limited military operations.

Although the Coastal Plain has no high hills or deep valleys, there are numerous swamps and rivers, the latter likely to flood in heavy rainfall. Such conditions made transporting armies about and keeping them provisioned demanding assignments indeed. Only once, in spring 1862, did the North's main army use the Tidewater for a thrust against Richmond. This was the Peninsula Campaign and it ended in near disaster for the Union. An especially rainy spring produced swollen streams and soaked ground that made fighting nightmarish for General George McClellan's attacking Army of the Potomac. Eventually, McClellan gave up the offensive and withdrew, defeated by Lee, the weather, and the topography.

The mountains of the Blue Ridge, western Valley and Ridge, and Plateaus acted as barriers because the steepness of the slopes was exceedingly inappropriate for offensive strikes. Traveling over broken, hilly terrain is physically taxing on men and animals, and keeping the soldiers together in coordinated attacks is an arduous task. No battles of decision were ever fought on the precipitous topography in these provinces. The infrequent gaps in the mountains, however, possessed much value for defenders because, as in previous wars, the passes "...allowed a few men to hold or delay large forces. Superior numbers cannot be brought to bear in a narrow defile" [172]. Confederate and Union armies often engaged in sharp fights for control of mountain passages, as at South Mountain and Crampton's Gap in the Blue Ridge when Lee invaded Maryland in 1862.

Unlike the barriers, the Piedmont and Great Valley corridors in Virginia saw nearly unceasing combat. The Piedmont is higher and drier than the low, wet Coastal Plain east of the Fall Line and much less hilly than the Blue Ridge to the west. Furthermore, the Union and Confederate capitals, both bound to be vigorously attacked and defended, were separated by only 100 miles of this firm, well-drained landscape. As a result, "…the country between the Potomac and the James became one continuous battlefield in the defensive and aggressive operations throughout the war" [210]. To quell the Rebellion, the Army of the Potomac had to take the offensive. Consequently, it repeatedly used the Piedmont terrain to march south from Washington and threaten Richmond. Though the land surface was good for battle, the Piedmont rivers, generally flowing from west to east between the enemy capitals, created significant stumbling blocks for the attacking forces. One of the biggest Piedmont streams—the Rappahannock—turned into the costliest barricade of all for the Union to overcome. Here Lee established his foremost defensive line, anchoring his army at Fredericksburg. It took the Army of the Potomac the better part of two years and tens of thousands of casualties in clashes at Fredericksburg (1862), Chancellorsville (1863), and the Wilderness (1864) to break through his staunch defenses.

The other prominent Virginia corridor was the Shenandoah Valley. This broad military highway functioned as a "…pistol in the hands of the Confederacy, pointed at the heart of the Union" [211]. (This evocative statement is Ellen Semple's rephrasing of Napoleon's comment that Antwerp, key to control of the Rhine River, was a pistol pointed at the heart of Europe.) Southern armies marching northeast down the Valley drew closer to Washington, Baltimore, and Philadelphia. But for the Northern forces, advancing to the southwest in the same feature moved them farther away from Richmond and deeper into the remote mountains of western Virginia. The open expanse of the Shenandoah is interrupted by a single imposing elevated feature—Massanutten Mountain, a fifty-mile-long sandstone ridge rising starkly nearly 3,000 feet above the valley around it. Generals "Stonewall" Jackson, Robert E. Lee, and Jubal Early all used the Shenandoah terrain to wreak havoc upon their opponents.

The first consequential military action in the Valley took place with Jackson's campaign in spring 1862. Using the Valley floor for fast-moving marches by his "Foot Cavalry" and Massanutten to mask those movements, the wily Stonewall's 17,000 soldiers fought five battles, several more skirmishes, and defeated three separate Union armies. This quick strike, hide-and-seek maneuvering ousted Federal units from the Shenandoah country. At the same time, the military masterpiece tied up 60,000 Northern troops, preventing them from joining the fight against Lee during McClellan's Peninsula Campaign.

Later in September 1862, Lee campaigned into Maryland by dividing his Army of Northern Virginia and advancing the two wings north along the Piedmont and Great Valley corridors. The Confederates reassembled at Antietam in the Valley of western Maryland to face McClellan's Army of the Potomac. The resulting battle was a tactical draw but a strategic Union triumph, for it forced Lee to retreat and end his invasion. His retiring army used the Valley route to reach the safety of Virginia. One year later, the Southern general again carried the war into the Northern homeland, this time marching his men along the Great Valley into Maryland and Pennsylvania where he met the Union army and defeat at Gettysburg. Once more, withdrawal back into the Shenandoah brought sanctuary and security for his vanquished warriors.

In 1864 Early assumed command of the Confederate units in the Valley. A series of battles broke out that summer when Early led his men north and reached the outskirts of Washington, causing much panic. (It was then that President Lincoln climbed atop the defensive earthworks to observe the Rebels in the distance and had to be pulled back to safety before a sharpshooter could take him down.) The Federals rushed in reinforcements that helped drive the Confederates back into Virginia. A new, aggressive Union commander named Philip Sheridan arrived in the Valley in August and fought a series of engagements with Early lasting into the fall of 1864 and spring of 1865. By the first weeks of March, Sheridan had eliminated the last resistance in the Shenandoah Valley; the now devastated region at last fell under complete Northern control 1 month prior to Appomattox.

Throughout the hostilities, the terrain in the Eastern Theater greatly affected battle at all levels. The Confederate generals often concealed their whereabouts behind the mountains in western Virginia and Maryland, at the same time employing the valleys between the ridges as natural avenues for deploying their troops. To the east in the Piedmont, the first two years of the conflict saw Lee and Jackson range over the gentle landscape with fast-moving flanking tactics that confused, frustrated, and defeated a series of Union opponents. But once Grant took charge in 1864, the Army of the Potomac refused to retreat. Instead, it kept pressure on Lee and stopped him "…from using geography to best advantage…The result was that Lee's initial stunning Napoleonic maneuvers were replaced by Grant's grinding war of attrition, and …the tide of battle in the east shifted irreversibly against the South" [273].

Topography and the War in Southwestern Virginia

The landforms of southwestern Virginia had done much to shape the culture of its inhabitants. The rough terrain created a distinctive group of hill folk with a sense of social, political, and economic separation from the plantation society far beyond their homelands. When the cannon at Fort Sumter blasted apart the final bonds between North and South in April, 1861, many of the Appalachian mountaineers showed limited (at best) allegiance to the Southern cause. Feeling no strong connection to the government in Richmond, groups of such men chose not to join the army, fighting instead in free-wheeling guerilla bands. These bushwhackers, as they were generally known, at times cooperated with the regular army or more often operated as they pleased. Formation of such unruly partisan gangs not only hindered recruitment into the Southern armies but also encouraged desertion from the regular forces. Moreover, the spread of irregular units fostered the rise of opposition Unionist companies, resulting in pro- and anti-Confederate squads. This all amounted to little more than simply clusters of outlaws roaming the hills to rob or kill citizens of either political view.

The northeast-southwest geographic alignment of the southern Appalachians affected another group resisting participation in the regular armies—deserters. The deep valleys and sheltering ridges "…provided a fairly direct path home as well as innumerable places to hide from regular troops or provost officials" [167]. This was true for both Southern runaways headed into the lower South or Northern shirkers going the opposite way. Southwest Virginia in particular, with its location relatively close to the main armies and broad Great Valley pathway, grew to be a kind of deserter's haven where lawlessness generated terror among the residents being preyed upon. To quieten the mayhem, the regular army occasionally sent in detachments that sometimes found themselves fighting pitched battles with the renegades. As the war wore on, loss of faith in the Confederacy and the increasing disorder in the region sparked stronger forms of resistance. By 1864, at least one insurrectionist organization appeared, dedicated to the overthrow of the Confederate government.

In southwestern Virginia, the Great Valley lowland is narrower and the ground more broken up than farther north, yet it still presented a corridor that exerted tremendous influence on Civil War military actions. Nearly all of the mineral works resided here as well as the Virginia and Tennessee Railroad. To get at these objectives, Northern expeditions had to invade from Kentucky or West Virginia until the last months of the hostilities when the Valley route opened up. This was not an easy task in that it meant advancing and withdrawing men, animals, and equipment directly across the general trend of the western Valley and Ridge mountains. The treacherous topography—vertiginous ridges, narrow valleys, and numerous streams—together with the inferior roads made it virtually impossible to supply an army of any real size. Furthermore, small contingents of Southern soldiers posted atop the ridges could effectively trap larger enemy groups in the constricted defiles below, then cut their lines of provision and starve them into submission. Mountainous terrain always favors the side that holds the hill tops, and in southwestern Virginia that was the Confederates defending their home area.

It is noteworthy that the first truly successful Union operation to actually reach and damage the railroad and mineral production facilities did not take place until near the close of the war. That raid worked well in part because the Federals could finally invade Southwest Virginia using the Great Valley corridor, a much easier route than crossing over the mountains. The key was the late 1863 capture of Knoxville, itself located in the Valley of East Tennessee. In December 1864 General George Stoneman led his army from Knoxville northeast along the limestone highway into Virginia. As they progressed, Stoneman's men tore up stretches of the railroad and decimated the salt and lead installations. Although not permanently shut down, the output of these two critical raw materials ceased for a time when the struggling Confederacy could ill afford any losses to its dwindling resources base.

Besides the mountains and Great Valley, another topographic feature of southwestern Virginia figured prominently in military operations—the New River. First encountered by white explorers in 1671, mystery and controversy have surrounded this stream since. Rising in western North Carolina, it is the only river to flow northwest directly across the grain of the Appalachians. (The New River is part of the upper reaches of the Mississippi-Ohio drainage system, ultimately emptying its waters into the Gulf of Mexico.) Owing to this cross-cutting path through the highlands, the New is considered to be one of the oldest streams in North America.

Not being able to flow uphill, rivers always course around mountains unless their paths had been well established before the higher elevations arose. This suggests that the primal New may have already been present approximately 350 million years ago when continental collision first began uplifting the Appalachian chain. With mountain building going on all around, the ancestral New managed to stay in its course to the northwest, cutting down through the ridges as they gained height.

Since the Late Paleozoic upheavals, the river has doggedly maintained that same path down the slopes of the western Appalachians toward the continental interior. In a sense, the modern river is no older than yesterday, because—like all watercourses—it constantly shifts position through erosion and deposition. Even so, if the river's fundamental location in southwestern Virginia and direction of flow were indeed fixed eons ago before the birth of the mountains, the superbly misnamed New must be among the most ancient streams in the world.

The unusual transverse path of the New River through the mountains takes it across the Great Valley, making it a significant obstacle to north-south travel along the limestone pathway. In the underdeveloped southwestern Virginia of the 1860s, few bridges existed across big streams like the New. Getting across such rivers had to be accomplished by boat, ferry, or wading at the few fords available. For armies, river crossings are always difficult maneuvers, particularly under fire. This being the case, military forces campaigning in Southwest Virginia tended to stay on one side or the other of the New.

For the Virginia and Tennessee Railroad builders, the presence of this waterway also posed a nettlesome topographic problem. The river had to be crossed at some point, and engineers laying out the rail line chose Central Depot as the one place to bridge the stream. When completed, the span was 730 feet long and covered by a wooden-sided structure with a tin roof. It stood as the longest bridge over the largest river on the entire Virginia and Tennessee line. If this impressive edifice were destroyed, the railroad would be completely severed, at least for a time.

The route of the Virginia and Tennessee that brought the rails to Central and the New River had been built down the Great Valley from the north. Although not as wide or smooth-floored as the Shenandoah, the Valley in southwestern Virginia is still by far the easiest path for any road or railroad following the trend of the Appalachians into the Deep South. Using this corridor placed the railway between the craggy Blue Ridge peaks to the east and the high-standing Alleghenies to the west, thereby avoiding numerous potential construction and maintenance issues in those more exacting landscapes.

Within the Valley, however, the railroad did have several intimidating terrain problems to overcome, for a number of rivers and one bold escarpment lay in its path. Just south of Big Lick (present-day city of Roanoke), the route ascended the eastern continental divide, an 800-foot change in altitude that separates two major drainage systems. To the north, the rivers are much lower in elevation and flow east into the Atlantic Ocean. South of the divide, the stream networks are at a higher altitude and drain into the Gulf of Mexico. Once the railroad tracks passed up and over the crest of the escarpment and entered the Gulf drainage reaches, the Virginia and Tennessee was in the New River Valley.

When the war got underway and cutting the Virginia and Tennessee became one of the main goals of Federal strategy, the so-called "Long Bridge" over the New River beckoned as the obvious target. Ripping apart a railroad for any considerable length is not feasible for the kind of relatively small raids that the Union commanders in the region could mount with the men and resources at their disposal. Therefore, obliterating the most vulnerable structure—the bridge at Central Depot—fixated Northern planning from the outset. And defenders of the Long Bridge would have their hands full throughout the conflict trying to prevent enemy raiders from doing exactly that.

Notes

The title of this chapter is taken from Clifford Dowdey's [48] book by the same name. The introductory account of how the geology of Gettysburg affected the fighting there goes back to Brown [26]. This author was a U.S. Geological Survey geologist who wrote several popular pieces on geology and the Civil War. I first read these as an undergraduate geology major when they were originally published in the early 1960s; his articles sparked my life-long interest in the linkages between geology and military history. The following section overviews the battles at Cloyds Mountain and Saltville and is condensed from McManus [144] and Marvel [131], respectively, with the effects of geology and terrain added from my work [258, 259]. The lengthy section on "Rocks and War" is deeply grounded in Semple [212], Chap. 14. The significance of the three great physiographic features of Virginia—the Chesapeake Bay, Piedmont rivers, and Shenandoah Valley—is her work entirely. I urge anyone interested in the linkage between terrain and Civil War strategy and tactics to read this geographer's insightful work. Harold Winters et al. [274] is the source of the corridors and barriers analysis; Zabecki [284] and O'Sullivan [173] provided some material on the importance of terrain in war also. The "Topography and the War in Southwestern Virginia" narrative begins with the impact of the mountains themselves on political attitudes, guerilla war, and desertion, a discussion taken mainly from Noe [169], and complemented by information from Walker [246] and Marvel [131]. The geologic history of the New River I found in Houser [86], Lowry [125], and Bartholomew and Mills [14]. Walker [246], Johnson [98], Marvel [131], McManus [144], and Noe [169] all stress the importance of the New River Bridge at Central Depot.

Chapter 5
Niter and Gunpowder

*"The entire supply of gunpowder ... was scarcely
sufficient for one month"*

On March 15, 1959, John Salling died in Slant, Virginia, his lifelong residence in
the far southwestern quarter of the state (Fig. 5.1). Salling claimed that he was born
in 1846 and served in the 25th Virginia regiment of the Confederate Army where he
was assigned to help dig out the local saltpeter deposits for the manufacture of gunpowder. At age 14 or 15, Salling said, he went to work collecting "peter dirt" in his
home area of Scott County for the army's Niter Corps. He told one reporter that he
never wore a service uniform: "There just warn't enough to go around" [57].
Although niter-rich limestone caverns known to have been active in the war abound
in this vicinity, Salling stated that he did not work in the caves. Instead, his unit "...
got dirt from the floors of chicken houses, tobacco barns, stables, and under houses"
[57]. Toward the end of his long life, he attended many veterans' meetings where
people who saw him invariably were struck by his thick shock of unruly dark hair
standing out amid his white-headed comrades. When asked how his hair had stayed
that way, he replied it was because he never touched it with water.

Over the years since his death, John Salling has been cited many times as the last
Confederate soldier to die. Much controversy revolves around who was in fact the
last Southern serviceman to pass on and Salling's actual age and record of enlist-
ment are certainly unclear. More than a few of his statements about himself and the
war years are contradictory. Regardless, the Veterans Administration eventually
decided to give him a monthly pension, and in 1956 he was awarded one of the four
medals the United States had prepared for Confederate veterans. In the following
year, President Eisenhower sent his personal greetings when the crusty former
Rebel received recognition as an Honorary Member of the Civil War Centennial
Commission. Members of the Women's Army Corps fabricated a special Confederate

© Springer International Publishing Switzerland 2015   47
R.C. Whisonant, *Arming the Confederacy*, DOI 10.1007/978-3-319-14508-2_5

Fig. 5.1 John Salling at the age of 108 years on his birthday, May 15, 1954. Photograph by G. Alex Robertson, from the Clay Perry Collection, courtesy of the National Speleological Society. According to some, "peter monkey" Salling was the last Confederate veteran to die

gray uniform and presented it to him. Two years before he died, the sturdy old Virginia "peter monkey" finally had his uniform. At his funeral, mourners commented on the luxuriant black thatch of hair still perched atop his head.

In the Civil War, the Confederacy's armed forces faced shortages of many critical materials, yet gunpowder rarely ranked among them. In the 1860s, potassium nitrate served as the principal ingredient in the black explosive mixture used widely by civilians for hunting game and blasting and by the military for more lethal purposes. Thanks to an abundance of saltpeter caves and unusually capable administrators, the South built a first-rate niter and gunpowder industry almost from the ground up. Even at the end of the hostilities, powder mills operated and stores of gunpowder were on hand. For centuries, purified potassium nitrate had been obtained from processing niter or saltpeter, as the source deposits were called. Each powder grain contained 75 per cent nitrate, 15 per cent charcoal, and ten per cent sulfur. When war came in 1861, the Confederates did not possess an adequate stock of gunpowder. Union forces captured a portion of the stores at Norfolk, leaving only low quality reserves from the Mexican War (1846–1848), in all approximately 60,000 pounds, remaining in the Southern arsenals. Planned importation of black powder could not meet all of the wartime needs, as the North's blockade of the ports soon proved.

Thus the demand for a strong, home-based gunpowder supply grounded in a reliable source of niter became evident. Government officials immediately looked to the numerous saltpeter caves in the limestone terrains of the Southern states as potential primary providers of potassium nitrate. Limestone is a key factor in cave occurrence due to naturally acidic ground water dissolving away the calcium carbonate composing the rock, thereby creating the underground chambers. Virginia has a proliferation of caverns with nitrate accumulations in the Great Valley and the other limey rock belts west of the Blue Ridge. Many such saltpeter caves had been discovered and mined for nitrate in the earliest frontier days before the Revolutionary War. In the Civil War, Virginia put out more of this absolutely essential war-making product than any other Confederate state, most of it coming from the grottoes in the western part of the state.

Niter and the Age of Gunpowder

Knowledge of niter extends back to the earliest times of recorded history. Sumerian writings from about 2200 BCE refer to both saltpeter and black saltpeter, the latter suggesting that refining of this material had already been accomplished. The Ebers papyrus, a scroll found in Egypt at Thebes and written circa 1550 BCE, mentions saltpeter and other resources obtained from the earth. The Old Testament Book of Leviticus, dated around 570 BCE, apparently refers to niter encrustations on the inside walls of Israelite houses (Leviticus 14:37). Alchemists in Europe knew of saltpeter in the first century BCE, probably by way of sources in China that traveled to the West through trade. By the seventh century CE, Chinese workers had mixed potassium nitrate with other ingredients to make fireworks, and in the tenth century described a military application of an explosive combination of niter, sulfur, carbon, and asphalt.

An Arab scientist in 1270 recorded the refining of saltpeter into concentrated potassium nitrate for gunpowder and explosives. In the Muslim world this material was called "Chinese Snow" or "Chinese Salt." Local sources of saltpeter came to light about this time in the Middle East, one being in a valley between Mount Sinai and Suez. During the Middle Ages, importation into Europe of niter from China and India continued, and the initial reports of saltpeter production in the West appeared. This early product was harvested from animal waste in farm outbuildings and stables.

In 1248 English scientist Roger Bacon became the first European to document the making of gunpowder, specifying its composition as saltpeter, sulfur, and charcoal. Gunpowder may have appeared on a European battlefield as early as that same year when the Bishop of Lyon reported the use of cannon by the Moors at the siege of Seville in Spain. Numerous instances of gunpowder firing projectiles from crude artillery in Western battles are chronicled in the early fourteenth century, for example in France at the Battle of Crecy in 1346. In these and later medieval times, the pace of work quickened on the collection, processing, and applications of purified potassium nitrate in warfare and other endeavors. Commercial uses included the production of metals, fabric dying, glass making, and preparation of various medications. The original modern book on mining and metallurgy, *De re metallica* published in 1556, discussed the extraction and refining of saltpeter in detail. Apparently, artificial niter beds had been developed by then as well. Within a few years, the rapid onset of the gunpowder age made niter an increasingly crucial resource in the countries of Europe, not only to supply the emerging national armies but also to ensure the survival of their colonists in the hostile New World.

Almost from the time of arrival in 1607 at Jamestown, settlers in America worried about a reliable source of saltpeter for their weapons. In 1629–1630, the Virginia colonial government passed an act "…for the better furtherance of and advancement of staple commodities, and more especially that of potashes and saltpeeter" [59]. This early legislation contained specific directions for getting niter from wood ashes and refuse from plants and animals. In 1745 the Virginia General Assembly approved an act for the encouragement of saltpeter making that offered a reward on the precious compound. The growing demand for niter made locating and mapping caverns that might contain nitrous earths a requirement. Among the earliest cave maps published in the United States is a 1770 drawing by Thomas Jefferson of a cavern in western Virginia that later yielded niter.

As war between England and her American colonies loomed in 1775, the Continental Congress advised the colonists to "…collect the saltpeter and sulfur in their respective colonies…to be manufactured, as soon as possible, into gunpowder" [60]. A national Committee on Saltpeter was formed, and Richard Henry Lee represented Virginia on this body; other members included Jefferson and Benjamin Franklin. By 1783, there were more than 50 caves with active niter mines in one Virginia county alone. Records from the Revolutionary War period are poor, but caverns in western Virginia, which then included today's state of West Virginia, likely provided a considerable amount.

After the surrender of Cornwallis at Yorktown, the demand for saltpeter did not abate. Indeed, frontier defense, hunting, government military uses, and the expanding application of black powder blasting in mining and construction drove the need upward. In addition, the War of 1812 contributed substantially to increased production in the early nineteenth century. In these times, more and more gunpowder-making facilities sprang up, owned and operated by private citizens digging saltpeter from limestone caverns. Because the limestone bedrock also forms valleys with fertile soils, such caves tended to be in lowlands already homesteaded, making them more likely to be found and exploited. By the mid-1800s the grottoes of western Virginia had a long history of nitrate productivity, including for two national wars. They would be called on again in the Civil War.

From Cave Earth to Explosives

The connection between caves and nitrate-rich deposits has been known and exploited for centuries. For most of this time, organic material, primarily bat droppings, was assumed to be the origin of the underground nitrates. More recently, different ideas have appeared that ascribe the formation of cave saltpeter to a complex interaction involving nitrogen-rich surface soils, groundwater, and nitrate-fixing bacteria. These studies compared niter caves in the southeastern United States with caves in the western part of the country where plentiful bat guano accumulations occur, but only relatively minor quantities of saltpeter. Even though bat excrement can enrich cave earth in nitrates, it is not a prime contributor in the southeastern caves. Here, bacteria in deciduous forest soils turn nitrogen into nitrate, which is then dissolved into the rainwater and snowmelt sinking into the subsurface. The infiltrating waters move toward underground openings, seeping into cave sediments where the bacteria work again, this time to precipitate the nitrates.

This theory explains a long known property of subterranean saltpeter occurrences—the nitrate content can be regenerated in very short spans of time. In one famous example, saltpeter miners in 1812 shoveled earth they had previously leached of nitrate back onto the wall ledges of Dixon Cave in Kentucky for the express purpose of creating fresh concentrations. In the Civil War, a saltpeter maker suggested that dirt be carried into caves so as to become continuously charged with nitrate. And, in fact, workmen often went back to previously depleted saltpeter deposits and found them once again replete with mineable accumulations.

Because conditions of soil, temperature, and moisture must be within fairly narrow limits to engender the resource, most American niter caves are located within the boundaries of the old Confederate states. Grottoes located in the North tend to be in zones where the temperatures are too low and too much water is present. Caverns can also be too far south where higher temperatures and soils containing less nitrogen inhibit nitrate formation. To the west, the drier climates and presence of soils that do not release their nitrogen readily are barriers to saltpeter development. In the 1860s, the South benefitted immeasurably from having the proper natural conditions that gave it the greatest concentration of niter caves in North America.

Transforming the raw cave sediments into niter involved a relatively simple process that could be done on site using widely available agricultural implements. Workmen dug out the nitrate-bearing earth using various tools like shovels, mattocks, and hoe-like scrapers. Additional equipment included iron pots, small tubs, a few water troughs, wheelbarrows, and wooden barrels. In big operations, mules, donkeys, or oxen hauled excavated cave earth to the processing sites, commonly located within the caverns themselves.

Leaching the peter dirt to concentrate the nitrates was the main refining process. Workers placed the cave diggings in vats or barrels, commonly three in number, containing fresh water (Fig. 5.2). Water leached from the first barrel was poured into the second and then into the third. Treatment of the nitrate-rich leach water with potassium salts obtained by soaking wood ashes followed. Next, workers removed more impurities from the leachate by straining it through cheesecloth, then boiled it in open iron kettles. Evaporation of the heated water left behind crystals of potassium nitrate; any remaining water got recycled back into the first vat to begin the purification process once again. Employing this technology, three men in Civil War times could turn out 100 to 200 pounds of saltpeter every three days (Fig. 5.3).

The ultimate destination of the crudely processed niter was the gunpowder mill where further refining of the saltpeter took place by additional leaching and boiling down the resulting concentrated liquid. Mill workers added sulfur and charcoal to the purified niter, making the compound highly explosive. At the Confederacy's greatest powder works in Augusta, Georgia, powder mixes in 60 pound batches used this standard recipe: 45 pounds nitrate, nine pounds charcoal, and six pounds sulfur. Final processing dampened the mixture and pressed it into solid cakes which

Fig. 5.2 Remains of a typical leaching vat used to extract nitrate from cave "peter dirt." In some instances, a series of such vats was used to progressively concentrate the leached saltpeter. Photograph courtesy of Ernst H. Kastning, Jr.

Fig. 5.3 Example of tally marks found in many of the caves worked for saltpeter in Virginia and elsewhere. Apparently used to record niter production, these marks are from a cave in the Shenandoah Valley. Note the name and date (Samuel Baker, 1862) inscribed on the cave wall. Photograph courtesy of Ernst H. Kastning, Jr.

were then cooled and broken up into grains. Vibrating wire screens separated the grains into different sizes—smaller ones for rifles and smoothbore muskets, larger ones for cannon.

Confederate Niter and Powder Manufacture

When war broke out, the Confederacy found itself in dire need of niter and gunpowder. True, numerous saltpeter caves throughout the Southern states had earlier been found and nitrate taken, but those sites for the most part stood idle by 1861. The long years of antebellum peace and emphasis on agriculture at the expense of industrial development resulted in reduced niter operations and only two active gunpowder-making factories. Colonel George Rains, chosen to direct the Confederate powder industry, put it bluntly: "The entire supply of gunpowder in the Confederacy at the beginning of the conflict was scarcely sufficient for one month..." [186]. A shortage persisted for the rest of that year, and only importation was able to provide most of the nitrate consumed by the domestic powder manufactories.

That same year the North undertook its blockade of Southern ports, an embargo that grew stronger and more effective as the fighting wore on. Confederate officials realized that smuggling niter past the Union ships could never completely meet the accelerating needs. With this in mind, the Richmond government placed tremendous emphasis on creating domestic sources of nitrate and powder mills to process it. Before long, niter manufacture soared, tripling in 1862 and again in 1863.

Although importation remained a major supplier of gunpowder throughout the war both by sea and overland from Mexico, Confederate administrators moved swiftly to bring about three home-based sources of saltpeter. First, as noted earlier, saltpeter caverns located mostly in the southeastern region had long been the basis of gunpowder manufacture in the United States. Many of these dormant operations were brought back on line, and the states of Virginia, Tennessee, Georgia, Alabama, and Arkansas emerged as the most important suppliers of cave niter for the Rebellion. Second, nitrates could be recovered from "dirt" under the houses and outbuildings of the citizenry. Mining these accumulations demanded the cooperation of private citizens, not always a stable piece of the supply chain.

A third and more reliable way to procure saltpeter, one utilized since the Middle Ages, was from "nitriaries" or artificial niter beds. Early in the hostilities, nitriaries composed of manure, rotted vegetable matter, and other kinds of human, animal, and vegetable waste were set up, usually near large cities. Workmen tended these "nitrate gardens" assiduously, moistening them with urine from time to time, then turning the decomposing compost over to generate a coating of nitrate. The war years saw at least 13 nitriaries established, including some near Richmond; at Selma and Mobile, Alabama; at Charleston, South Carolina; and at Savannah and Augusta, Georgia. The artificial beds, however, needed months of cultivation before generating usable quantities of saltpeter. In the end, most had not had enough time to yield significant nitrate crops prior to the close of combat in 1865. Still, estimates are that three to four million pounds of niter would have been produced from this method, a substantial volume considering that a total of about three and a half million pounds of nitrate was turned out over the course of the conflict.

Starting with almost nothing to build on, the Confederacy's niter and gunpowder manufacture succeeded spectacularly, thanks mainly to the selection of three extremely able leaders—Josiah Gorgas, George Rains, and Isaac St. John. Gorgas was a West Point Military Academy graduate and Mexican War veteran with many years of experience in munitions production. His service in the United States Army included stints at several arsenals dispersed across the nation. During those tours, he worked on new types of cartridge and artillery designs, and became well versed in the preparation of gunpowder. In 1861 the Confederate government picked Gorgas, then a major but promoted to Brigadier General three years later, to lead the Ordnance Department. He promptly set to work building an armaments industry. Grasping the strategic necessity of the raw material required for gunpowder manufacture, he suggested in 1862 that a Niter Corps be formed. He nationalized the saltpeter procurement effort by convincing the individual states to give up their holdings to the Confederate government. Gorgas eventually saw the need for a Niter and Mining Bureau separate from the Ordnance Department and successfully lobbied the Confederate Congress for its creation in 1863.

Shortly after the war began, Ordnance Chief Gorgas chose Colonel George Rains, another West Point graduate who later taught chemistry, geology, and mineralogy at the Academy, to oversee the gunpowder-making industry. This talented administrator started his work with virtually no infrastructure; two little powder mills in Tennessee and South Carolina were all that existed. Soon Rains' determined efforts and creative ideas had much larger and more efficient gunpowder plants coming on line, including an innovative facility at Augusta that later on became the second biggest powder factory in the world. In addition to establishing a national gunpowder production system, Colonel Rains strove to ensure a steady supply of niter for his factories. He and his assistants visited niter caves in Georgia, Alabama, Tennessee, and Arkansas, identified those with the best potential, and signed contracts with the owners and operators to provide this essential material.

The energetic Rains' masterwork was the giant Augusta Powder Works, a technologically advanced operation that manufactured gunpowder for the Confederate Army equal in quality to the finest European output. (The Navy had its own powder mill at Columbia, South Carolina.) A small Federal arsenal had been in Augusta for years, and Rains saw many advantages to expanding this site. Water and railroad transportation lay close at hand, as well as abundant high-grade wood for charcoal. The mild climate afforded relatively comfortable working conditions year round, and Augusta's location in the Southern interior reduced the likelihood of enemy assault. To upgrade the security of the site, Rains saw to the construction of earthen forts surrounding the city and organization of a local defense force. He also built a test facility south of town where 12-pound Napoleon cannon cast at the arsenal fired shots across the river using Rains' explosives.

Erection of the Augusta Powder Works began in September 1861 and the plant opened the next April. Much of the heavy iron equipment needed for powder manufacture came from the Tredegar Works in Richmond, the only facility south of the Potomac River that could make such items. Rains devised a variety of improvements in the processing, including enhanced purification of the saltpeter and sulfur going into the powder. Besides the grade of the explosive, he fretted about accident prevention. He insisted that the main buildings be separated by at least 1,000 yards and set up tree and brush barricades to reduce damage from flying objects should a blast occur. His engineers redesigned work spaces to be flimsy and easily blown apart in those places where a mishap was most likely to occur. Four explosions did happen over the course of the war, yet the Augusta arsenal never shut down due to accidents, raids, or the ever-worsening labor shortages.

When the Richmond government established the Niter Corps in 1862, Gorgas selected Lieutenant Colonel Isaac M. St. John to lead the unit. A year later the corps evolved into the Niter and Mining Bureau and St. John headed this agency until peace arrived in 1865. Like Gorgas and Rains, St. John proved to be an immensely gifted administrator, said after the war to have "...created resources where none previously existed" [178]. After graduation from Yale, he worked as a lawyer, newspaper editor, and civil engineer ,employed by a number of railroad companies. Beginning as a private in a South Carolina regiment, his accomplishments in providing gunpowder for the Southern armies earned him promotion to brigadier general in 1865. As Bureau chief, he directly supervised the collection and production of saltpeter from all sources as well as procurement of the other minerals within his agency's responsibility. St. John organized the Confederacy into 14 districts clustered into three divisions. This improved efficiency and his smoothly running Bureau achieved much with remarkably few officials and laborers downstream.

St. John struggled constantly with a lack of workers. The Confederate government had exempted workmen employed by the Bureau from military duty (this being about the only way to keep them). On the other hand, they were still subject to military discipline and could be used to repel Union incursions. Some of the labor difficulties rooted directly in the tepid support for the Confederacy or outright Unionist sympathy typical of the mountain people living in the nitrate cave regions. These attitudes resulted in a notoriously unreliable workforce where absenteeism and desertion were rampant, and far fewer slaves were available in the hill country to make up the losses. Because many of the other vital mineral industries operated in the same highlands, such problems plagued mining efforts throughout the war.

After a time, the Niter and Mining Bureau gained the authority to enlist workers from those men subject to military service. In spite of the worsening manpower conditions for the South as the fighting raged on, the Bureau managed to augment the staff laboring for it. Niter works controlled by the agency had 2,771 whites and 115 blacks in its employ by July 1862. (These figures do not include the collection and processing performed by private citizens.) The Bureau had the right to impress slaves into its ranks, but it also hired many free blacks.

St. John originated training programs wherein a core of skilled laborers mentored those just coming into the operations. The Bureau chief firmly believed that better training was a prime factor in making up the losses of niter workers. The Army resisted the Bureau's taking of men who otherwise would have been available to fight; St. John persevered, however, and by 1863 his agency had expanded its labor force to around 4,000 workers. Even though this total ultimately declined, the training of the workmen and the efficiency of the Bureau continued to push up niter yields until the last months of the contest.

With only a few thousand niter workmen available, the Niter and Mining Bureau generated most of the saltpeter supply. In contrast, the North used over 80,000 men for its niter production. Under managers like St. John, Gorgas, and Rains, Confederate niter and gunpowder manufacture expanded enormously during the conflict. Near the end in 1865, output of niter and other necessary raw materials diminished due to loss of geographic areas, a growing frequency of raids, and the unavailability of labor. Nonetheless, the leadership of the Bureau had worked wonders to keep things going as well as they did.

Virginia's Grottoes of War

Within the Niter and Mining Bureau structure, Districts One, Two, Three, Four, and Four and a Half resided in Virginia, and included the saltpeter producing counties in present-day eastern West Virginia. Surviving documents show niter output to September 1864 totaling 1,735,532 pounds domestically and 1,720,072 from imports for the entire Confederacy. Of the internal manufacture, the five districts in Virginia accounted for 505,584 pounds (about 29 per cent), making it the leading nitrate provider among all Confederate states.

No one knows precisely how much of the Virginia contribution came from caves or how many caverns turned out niter in the war. Records were lost or poorly kept (if at all), particularly by the scattered private contractors who sold directly to the government. Another confusing issue is that cave names have changed or varying local names have been used for the same site over the years. Saltpeter digging and processing occurred at approximately 100 Virginia caverns over the state's centuries-long history of production, and a considerable fraction of these must have been active in 1861–1865. The best estimates are that about 25 caves in Virginia manufactured niter in that time. A July 1, 1863, report from Niter and Mining Bureau Director St. John to Confederate officials concerning saltpeter operations in western Virginia described the situation:

> In Virginia, contracts have been closed with some fifty parties, some of whom have worked well, but from their slow progress and the frequent loss of caves by the enemy in Greenbriar and Monroe Counties, work on Government [account] has become a necessity as follows; one large cave in Tazewell, one in Giles, and six small caves in Wythe, Smythe [sic], Pulaski, and Montgomery. These caves are in good working condition and are beginning to yield... [61].

What were the day-to-day activities like at a wartime saltpeter cave in western Virginia? The tools and other artifacts left behind by the peter monkeys tell part of the story, but detailed accounts of the daily work routine are not plentiful. Still, in a few instances good descriptions of the cave operations can be pieced together, three examples being Buchanan Saltpeter Cave in Smyth County, and Horner's and Heaton's niter caverns in Bath County.

Buchanan Saltpeter Cave is located in the Great Valley in northern Smyth County. These caverns have an extraordinarily lengthy history of being mined for saltpeter, one that involves connections to another famous mineral resource in the locality— the salt works at Saltville. In 1748, John Buchanan belonged to a party of men sur- veying property along the North Holston River for a real estate development company. The group included Charles Campbell, who recognized the value of the salt brines they discovered and acquired the site that at first was known as Campbell's Salines and later became Saltville. Soon after settling there, Campbell had salt making underway. Meanwhile, Buchanan surveyed and claimed broad holdings in nearby Rich Valley, then established his home there and set about farming.

Gunpowder was a necessity on the frontier and an extensive cave existed on Buchanan's property. Aware of the grotto and understanding that such places com- monly contained nitrate accumulations, he investigated and found considerable deposits of peter dirt. The entrepreneurial Buchanan commenced saltpeter extrac- tion but needed some of the heavy iron boiling kettles frequently used to concen- trate the nitrate. Charles Campbell employed these very kinds of kettles in his salt making activities; before long, the two businesses cooperated by sharing the big iron vats and other equipment. J. Leander Bishop, a nineteenth century medical doctor and historian from Pennsylvania, wrote about American industry during the late 1700s and noted the following about the growing Smyth County mineral enter- prises: "Salt was made by boiling at Campbell's Salines… and in 1795, several tons of saltpeter, collected from the nitrous caves in the county, [and processed at the salt works] were sent to the Atlantic market" [62]. Considering that Buchanan's business constituted the only major source of cave peter dirt in the county, it almost certainly had to be the source of the Saltville niter.

Buchanan Cave rated as one of the primary western Virginia saltpeter suppliers in the Civil War. Long experience in niter manufacture had sharpened operational efficiency considerably. A clever cascade system evolved that used a series of wooden leaching vats arranged so that the leach water from one would drain into the next lower one. This gave advantages over single vat techniques, including less water needed, improved leach concentrations, and reduced loads of fuel (and the labor to cut the wood) required to evaporate the leach waters from the boiling ket- tles. Although no records survive to document the quantity of saltpeter from Buchanan Cave, the size and duration of the business indicate that it must have been a copious amount.

Horner's and Heaton's niter-producing facilities in Bath County lay within Niter and Mining Bureau District Four and a Half, administered from Staunton. Five known Confederate government sites functioned within the bounds of this district, four of which were in caves, including Horner's and Heaton's. Robert L. Horner, a farmer who had served in the Confederate army, earned $75 per month from the Bureau as superintendent for his site. He ran a sizeable operation, employing

70 different workers between November 1862 and July 1864. Of these, 17 are listed as "free Negroes" who toiled along with an additional three "black hands," apparently slaves. For any one month from January 1863 to October 1864, the number of laborers varied from 18 to 45. The pay per day totaled 60 cents and desertions—sometimes called "French Leave"—happened routinely. Records suggest that some makeshift housing was available to the laborers and purchases of food to feed them included flour, beef, and bacon. At least two horses and one or more wagons served the daily work effort. Horner's operation sent 89,724 pounds of saltpeter to the niter refinery in Lynchburg in the second quarter of 1864.

An ex-Confederate infantry captain named Nathaniel Heaton owned and managed a second niter works in Bath County. Heaton became an agent or assistant superintendent in Niter District Four and a Half in December 1862, earning him an extra $100 each month. Working conditions and the headcount of laborers closely resembled those at Horner's location. At Heaton's works, payroll lists record 60 different men from April-June 1863 to October 1864. At that time, the number of free workmen ranged from 17 to 35, and slaves toiled here as well. Desertions at Heaton's occurred less frequently than at Horner's, but other aspects of everyday life seem similar. Saltpeter from Heaton amounted to about 50 pounds daily for May 1863, 3,019 pounds for the second quarter of 1864, and 1,866 pounds in October 1864.

Threatened or actual Federal raids disrupted niter making in Bath and neighboring counties at least four times. General William Averell's mounted troops accounted for three of these occasions in August, November, and December 1863. In the first assault, Averell's soldiers destroyed a few saltpeter operations in the area, but there is no mention of an attack on Heaton's or Horner's facilities. In November, after the Union victory at Droop Mountain in nearby West Virginia, a company of mounted infantry extensively damaged Heaton's works; nevertheless, the owner and his workers returned forthwith and made repairs. Averell's riders found the place once again in December, this time causing "...the destruction of the cave..." [217], according to a Confederate officer. Several of Heaton's white laborers were taken prisoner along with five slaves working for the Niter Bureau.

In June 1864, Union General David Hunter led his men into the region, causing a fourth interruption in saltpeter making. Sowing destruction in his path, Hunter drew his foes into a pitched battle at Piedmont that ended with the defeated Confederates flying from the field. The captured Southerners included Lieutenant Robert L. Horner, and Captain Heaton may have fought there as well. Not long after this engagement, Heaton took command of the local niter force upon its reorganization into one company.

For most of the conflict, the grottoes of Virginia and elsewhere in the South contributed the bulk of saltpeter feeding the Confederacy's surprisingly successful powder industry. The dispersed nature of the caverns installations and their location in remote areas kept them relatively safe from disturbance. In addition to providing a steady source, the quality of Southern cave niter was exceedingly high and that from Virginia ranked with the very best. Colonel St. John reported that niter from the state's caverns possessed a superior character and could be quickly refined.

As a result, the gunpowder derived from such outstanding domestic saltpeter was quite good, equal in grade to any that could be imported from abroad.

By 1864, as Confederate cave territories increasingly came under Union control, output from other nitrate sources in the Confederacy at last exceeded that from the underground chambers. In 1865 total niter and gunpowder manufacture declined, yet at war's conclusion the Augusta arsenal had 70,000 pounds of powder on hand, thanks in no small part to the limestone caverns in the Appalachian highlands. In all, Augusta manufactured 2,750,000 pounds of gunpowder, over half of the entire consumption by Rebel armed forces. The massive volume and consistent excellence of the Augusta output earned it many accolades—The London Times printed articles lauding the factory and the powder—but perhaps the Union gave Colonel Rains the finest tribute of all. When hostilities ended, the United States Artillery School used the leftover Augusta powder for gunnery practice at Fortress Monroe owing to its superb quality.

Notes

The information on John Salling that begins the chapter is taken mostly from Faust [58] with some input from Hauer [76]. The overview of Confederate niter that follows is based on Schroeder-Lein [204]. In the "Niter and the Age of Gunpowder" section, Lewis [118] is the source of the history of gunpowder and niter from ancient times to the 1600s. Faust [63] provided copious information on saltpeter in early America and Virginia specifically; Powers [179] was used also. Concerning the "From Cave Earth to Explosives" material, Hill's [82] article is the basis of the discussion about the origin of saltpeter in the caves of the southeastern United States. For this topic, I also took information from Hubbard et al. [89] on the chemistry and mineralogy of cave saltpeter. Several articles, primarily Faust [63], De Paepe [43], and Powers [179] provided information concerning mining and processing of niter from caverns. In the "Confederate Niter and Powder Manufacture" section, the general discussion of the topic is based on Powers [179], Schroeder-Lein [203], and Lynch [127]. The lengthy look at the masterminds of the Southern niter and gunpowder effort—Gorgas, Rains, and St. John—is from Rains [188], Sipe [214], and Melton [150]. The concluding discussion of "Virginia's Grottoes of War" is abstracted from Faust [63] (Buchanan Cave), Smith [218] (Horner's and Heaton's works), and Hubbard [88].

Chapter 6
Bullets, Firearms, and Colonel Chiswell's Mines

"Discovering, raising, washing, and smelting the lead mineral"

Thomas Jackson was a determined man. Not wanting to miss his chance to purchase the Lead Mines in the backcountry of southwestern Virginia, the British immigrant traveled from Wythe County some 250 miles to Richmond where the state offered the metal works at auction. In the Virginia of 1806, this was a difficult journey. Local lore claims that Jackson made it without frequenting the bars and taverns along the way as did his competitors. Arriving at the sale first, his bid secured the Wythe County lead enterprise for himself and two partners.

These mines had been opened in the mid-eighteenth century by Colonel John Chiswell, a veteran of the British army seeking his fortune on the western Virginia frontier. After Chiswell died, later owners brought in Jackson to work at the mines in the late 1780s because of his extensive knowledge of mining and smelting gained back in England. Once he and his partners bought the mines outright at the Richmond sale, the hard-working Jackson took the helm as both owner and operator. Under his inspired leadership, the lead works expanded significantly over the next two decades. The mines turned out lead ammunition of various shot sizes for the growing settlements in the region and beyond. Jackson died in 1824, yet the firm footing he built for the mining operations carried them through the next several decades.

When the Civil War began, the mines in Wythe County immediately became essential to the Southern military effort, but not for lead shot. By then, large caliber lead bullets, in particular the Minie ball, had been developed for the rifled muskets that were the basic infantry weapon during the fighting (Figs. 6.1 and 6.2). The Minie ball actually had a bullet shape with a conical nose and a flat base, and measured over half an inch in diameter. These projectiles did frightful damage to the human body. The soft lead tended to deform on impact whereupon the jagged but still dense metal clump tore on into the soldiers, leaving gaping holes in tissue or shattered bone.

© Springer International Publishing Switzerland 2015

R.C. Whisonant, *Arming the Confederacy*, DOI 10.1007/978-3-319-14508-2_6

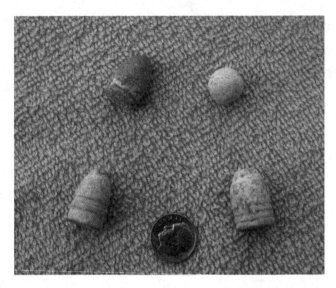

Fig. 6.1 Minie bullets (lower part and upper left of photograph) and buckshot (upper right). Lead projectiles such as these caused terrible wounds to Civil War soldiers. Size of the coin is 0.7 in. (13 mm.)

Fig. 6.2 Alabama infantryman with his rifled musket. Photograph courtesy of the Library of Congress. Development of the Minie bullet led to widespread use of these deadly weapons in the Civil War

Both North and South used immense quantities of the new cartridges, and "… the peculiar whistle made by the Minie bullet became unpleasantly familiar to many Americans" [31]. The clash of the Union and Confederacy would be the last big war where soldiers used simple unjacketed lead bullets. But in that conflict, the contending sides could not put armies in the field without an abundant lead supply to provide ammunition for the troops. More than any other single point of origin, the Wythe County mines provided the lead that poured from the rifles of Confederate infantrymen on Civil War battlefields.

A Brief History of Lead

Lead and humans have a long history together, often with deleterious consequences for the people exposed to this common base metal. Lead beads estimated to be about 9,000 years old have been found in an ancient Hittite city in Turkey. The low melting temperature of this element made it one of the first metals to be smelted, which could easily be done by placing galena, the ore mineral of lead, in open wood fires. Because galena almost always contains some silver, extractive measures such as the camp fire also created a small amount of this precious metal. For millennia, mines yielding lead were more valuable and better known for the silver by-product. For example, the famous mines at Laurion near Athens whose silver enabled the Greek naval victory over the Persian fleet at Salamis in 480 BCE produced lead as well.

For military purposes, lead's softness precluded any usefulness for the pointed or edged weapons of the ancient world, yet the high density made it an excellent ballistic projectile for the slings of the warriors. Furthermore, the malleability of the metal rendered it readily workable into different shapes. This property, plus its ease of smelting, widespread occurrence, and resistance to corrosion gave it value and common use in the early civilizations of the Mediterranean and Middle Eastern regions. As early as 4000 BCE, Egyptians employed lead as weights, pottery glazes, ornaments, and in eye-paint for facial makeup. The Israelites knew of lead, frequently referring to it in the Old Testament. Exodus 15:10, for example, records that the army of Pharaoh "sank like lead" when the sea water rushed in upon it. The Chinese used the metal for currency and the pre-Columbian people in Mexico manufactured lead amulets.

Around 2000 BCE, the Phoenicians, a sea-going people from the Middle East coast, extracted lead (and silver) from deposits in Spain. The Romans took over these prolific operations when they captured and destroyed the Phoenician-founded city of Carthage in North Africa in 146 BCE. Roman lead mines operated in Central Europe, Britain, the Balkans, Greece, the Middle East, and the Iberian Peninsula as the Imperial Age came into full flower. Rome consumed lead in astonishing amounts, raising lead output levels well beyond anything achieved before and not reached again until the Industrial Revolution 20 centuries later.

Lead was everywhere in the Mediterranean world—in water pipes, aqueducts, holding tanks, pottery and dishes, bath linings, coins, statues, pigments, and cosmetics.

The Roman word for lead was *plumbum*, and the metal's extensive use in pipes, many of them stamped with the emperor's image, gives us the English word plumbing. The cooking implements contained lead and the recipe books recommended it as a food additive. In the entertainment sphere, gladiators put lead coverings on their knuckles for combat in the arena. Wine held enormous importance in the days of empire, and its makers insisted on lead use throughout the process. The metal occurred in the processing vats, the decanters, and the drinking cups of the upper classes. (Poorer people had to make do with ordinary clay vessels.) In addition, lead has a slightly sweet natural taste, and vintners added it directly to the wine to improve flavor. It is not surprising that a number of Roman writers described diseases among metal workers and many other citizens that are obviously the effects of lead toxicity.

The production of silver and lead went hand-in-hand from the same mines and smelters as noted above. Silver taken from constantly worked and fiercely defended operations scattered about the Empire financed much of Rome's military might and high standard of living. At the same time, extracting and refining that metal meant that huge quantities of lead would also be available. In one of history's great ironies, the price of obtaining the precious metal to pay for Roman dominion may have included its eventual destruction, if the scholars are correct who say that lead poisoning, most concentrated among the aristocracy, hastened the fall of Rome.

With the collapse of the old imperium, demand for lead dropped steeply in Europe and stayed low for several hundred years. By the arrival of the tenth century new mines opened up, notably in eastern Germany. Thirteenth century English documents again mention widespread lead use. As medieval times neared their end, lead found increasing utility in roofing materials, coffins, water tanks, guttering, and personal decorations. In some places wine makers still routinely added the metal to their products or at least used it in the processing equipment and containers. When Johannes Gutenberg invented his printing press in the mid-1400s, lead gained a novel application as moveable type in the machine. But a far more lethal role for lead in world affairs had appeared by then, one fundamental to a revolutionary way of warfare—the Age of Gunpowder had dawned. And the murderously effective firearms coming onto the battlefield demanded cheap, easy-to-make projectiles.

Firearms and Bullets

The progenitor of all portable firearms was a crude kind of hand cannon; one is shown in use by a Chinese warrior in a twelfth century painting. These unreliable devices were difficult to handle and as dangerous to the user as to his enemy. Nevertheless, the weapons spread west and were known in Europe by the fourteenth century. Over the next 100 years, an improved gun called the culverin replaced the hand cannon (Fig. 6.3). This firearm had a long musket-like barrel attached to a wooden extension held under the arm. A gunpowder charge, ignited by a match touched to a hole at the base of the barrel, fired a round lead missile.

Fig. 6.3 Fifteenth century culveriners. The culverin was eventually replaced by the arquebus. Image from Wikimedia Commons

Late in the fifteenth century, an advanced tool of war, the arquebus, appeared in large numbers for the first time with infantry fighting in the area of Hungary (Fig. 6.4). This heavy and cumbersome instrument also required a match to set off the powder charge, had a slow rate of fire, and sent forth a low velocity projectile. Even so, arquebus technology changed warfare irrevocably, for the metal ball blasted from its barrel could penetrate heavy steel plate at close range. Now, with minimal instruction and a single shot from his terrible machine, the lowly foot soldier could bring down the lordly knight, clad in expensive plate metal and requiring years of preparation to become a skilled fighter on horseback. The day of battlefield domination by armored combatants thundering about on their equally armored chargers had ended. Gradually taking their place would be national armies composed of masses of infantry easy to train and inexpensive to equip with a more powerful killing weapon. With some assistance from cavalry and artillery, gun-toting infantrymen in the future would decide the issue on the field.

European armies swiftly adopted the arquebus and used it for the next 200 years. This fearsome tool of combat brought victory on the battlefields of Europe, the

Fig. 6.4 Late sixteenth century arquebusier. These shoulder arms revolutionized warfare as volley-firing infantry took control of the battlefield. Image from Wikimedia Commons

Orient, and other parts of the world. Portuguese and Spanish conquerors employed the arquebus in the sixteenth century to subdue the peoples of the New World and obtain its treasures for the Old. Around 1600, firearms evolution took another major step forward with the appearance of the flintlock musket. This sturdy and dependable long gun replaced the arquebus by the middle of the seventeenth century and stood preeminent in the global arsenals for the next two centuries. It was the smoothbore flintlock musket that American colonists shouldered to win their independence from England and afterwards to push the frontier farther west against resistance from the native peoples. Napoleon's troops triumphed in Europe with the flintlock, as did many armies around the world in the first half of the nineteenth century.

In spite of this success and constant improvements, the smoothbore flintlock could not overcome two persistent problems: its basic inaccuracy beyond more than a few tens of yards distance and slow rate of fire. In 1847, a French army officer, Captain Claude-Etienne Minie, introduced a distinctive lead bullet that transformed firearm technology. The Minie ball was the first projectile to effectively grip the rifling or spiral grooves cut into a musket barrel. This spinning, aerodynamically shaped missile discharged by a rifle dramatically improved marksmanship and rate of fire. First used in large numbers by the British against Russian troops in the

Crimean War (1853–1856), the Minie bullet led directly to the horrific totals of killed and maimed men on American Civil War battlefields.

Gunsmiths had known since the sixteenth century that cutting grooves into a musket and cannon bore imparted a spin to the projectile fired, thereby enhancing range and accuracy. The rifled musket, however, had not achieved widespread adoption for military purposes because powder buildup in the barrel grooves clogged the weapon badly and the extra time taken to remove it slowed the rate of fire considerably. Moreover, muzzle-loading arms like the musket required a ramrod to push in the powder charge and seat the bullet. The extra tight fit needed for a bullet to properly grip the rifling meant more time spent ramming in the projectile. With the need to load and fire rapidly being of overriding importance, loosely fitting lead balls easily pushed down a smooth barrel held sway as the standard technology in the flintlock era. For many years, therefore, musket balls flying from smoothbore barrels continued to careen uncertainly along erratic paths, much like a fluttering knuckle ball (a "dead" ball with no spin on it) thrown by a baseball pitcher.

The outcome of all this was that the most widely used infantry firearm for centuries was accurate up to about 50 yards; at 100 yards a man-sized object had a fair chance of being missed; and at 200 yards such a target was unlikely to be struck except by a stray bullet. As late as the 1840s, American infantrymen fighting in the Mexican War carried the same basic smoothbore weapon their ancestors had in the Revolutionary War in the 1770s and 1780s. And their brothers in arms in other nations were equipped with the same firearm.

The mid- to -long range inaccuracy of the smoothbore musket gave rise to offensive infantry tactics that became known as Napoleonic in recognition of refinements made by that master of warfare. This method of handling masses of soldiers had them march forward slowly in shoulder-to-shoulder ranks, pausing occasionally to fire in volleys at the enemy. The attack finished off with a bayonet charge if the defensive lines remained unbroken. Even though casualties could be considerable, military leaders of the times deemed them acceptable as the cost of battle. Officer training academies across the globe, including West Point in the United States, taught such maneuvers up to the Civil War.

The men destined to lead troops on both sides in those hostilities knew such tactics well, and had practiced them during the Mexican War. But the struggle between North and South would be very different. It was "…the first great conflict in which the combination of an accurate weapon and marksmanship of a high order forced radical changes in tactical formations" [31]. Nevertheless, those changes did not come quickly and unprecedented slaughters at places like Gettysburg, Fredericksburg, and Shiloh inevitably resulted from commanders clinging to Napoleonic formations. The tried and true methods of assault in the past simply did not work against the new technology now in the hands of the soldiers.

From the sixteenth to the nineteenth century, improvements made to muzzle-loaded firearms held against the shoulder or in the hand proceeded steadily, yet those same three centuries saw practically no innovation in the missile coming from their barrels. From the beginning, round balls of lead—a metal that was inexpensive, plentiful, and easy to cast into spherical shapes—was the projectile of choice.

(The French termed the little metal balls *boulettes*, from which we get the word "bullet.") The lead orbs fit smoothly into the barrels of the smoothbores, and development of a more aerodynamic form to improve range and accuracy was slow.

Early experiments on shape, including cubical, had not been fruitful, and it took until the 1820s for the first bullets with conical shapes to appear. These projectiles had a pointed forward end and worked fairly well; still, they did not engage the rifle grooves as they should for best effect and were not widely adopted. Ordnance officers, a conservative lot, showed little enthusiasm for further testing. The breakthrough emerged in 1847 when Captain Minie, building on earlier designs, created a truly practical lead bullet for rifles. The new missile slipped easily down a rifled musket barrel but had a special base that expanded upon firing, causing it to grip the spiral grooves effectively. Now, a soldier with rifled musket firing a Minie bullet could routinely hit a man 500 yards distant, and sharpshooters were better than that.

Almost overnight, military forces worldwide adopted the Minie bullet, more commonly called the Minie ball from the long years of lead spheres serving as the standard musket projectile. In the American Civil War, both Union and Confederate infantry fought with rifles shooting Minie balls almost exclusively. Estimates are that 90 per cent of the casualties came from such weapons. Unfortunately for the men under them, their leaders had not adjusted to this dramatic change in weaponry, and casualties among the advancing masses of troops packed tightly together skyrocketed. In one terrible day alone—September 17, 1862, at Antietam—23,000 soldiers were killed, wounded, or declared missing, mostly victims of the hail of lead tearing into them. The Confederacy manufactured an estimated 150 million bullets for the war, many of them Minie balls made of lead from the mines in Southwest Virginia. Metal from that source must have been flying into the Federal ranks at Antietam and in most every other battle fought between Blue and Gray.

The Wythe County Lead Mines

Though deposits of lead have been discovered in many places within the United States, the largest by far is the Tri-State District of Kansas, Oklahoma, and Missouri. In the latter state, the first lead mine in the country opened in 1720. In the eastern colonies of Connecticut and Massachusetts, miners found and extracted the metal sporadically over the decades of the eighteenth century. In southwestern Virginia, a small lead works had started up along the New River by 1761, ultimately growing larger and functioning continuously for the next 220 years. From the 1760s to the 1860s, the Wythe County mines supplied ammunition for armies engaged in three national struggles. In the first two wars, England was the foe; in the last, however, the enemy was the United States itself.

The lead deposits in southwestern Virginia occur in the Shady Formation, truly ancient rocks dating back to the early Cambrian Period 540 million years ago. At that time, long before the Appalachian Mountains arose, what is now eastern

North America was a low-lying coastal plain bordering a tropical sea. Along that shoreline, the Shady formed as layers of limestone that later turned into dolomite. The transformation into dolomite is a complicated chemical process wherein magnesium replaces some of the calcium in the limestone. This process takes place today along tropical coastal zones, such as the shallow waters and beaches of the Persian Gulf and Bahama Islands.

Emplacement of the lead ores into the Shady did not happen until much later when hot fluids rich in lead and zinc flowed through the buried formation, leaving behind rich troves of the metals. Most geologists who have studied these strata think that the metal-bearing waters came about when rifting or mountain building heated and deformed the continental margin. For the next several hundred million years, the Shady lead and zinc lay undisturbed until a former British army officer happened upon them.

Colonel John Chiswell, a native of Wales, immigrated to America and lived for a time in eastern Virginia where he engaged in a variety of enterprises involving mining and metallurgy. In the 1750s this entrepreneur was looking for economic opportunities in Southwest Virginia, then wild frontier country with white settlers just beginning to arrive. Destiny smiled on his prospecting efforts when he located one of the biggest concentrations of lead ever found in the eastern United States. The story of the discovery, perhaps legendary, is that Chiswell took refuge in a cave along the New River in the present-day Wythe County area while being pursued by Indians. Waiting for the danger to pass, he noticed shiny crystals of galena in the cavern walls around him and recognized the ore mineral of lead. Commencing in 1761, he had mines and smelters churning out the base metal, and before long saw the need for someone with abilities beyond his to enhance the extraction and processing of the raw ore.

Chiswell sailed to England in 1763 and returned with William Herbert, an expert in refining ores of metals, to manage his mines; some other Welsh miners and their families came along as well. By the fall of that year, wagon loads of Chiswell lead were being bartered for supplies in North Carolina. The lead works prospered for the next few years, then tragedy struck John Chiswell in 1766. He was accused of murder in a sensational case that stayed in the Virginia newspapers for weeks. Traveling to Williamsburg, Chiswell got into an argument with a Scot in a tavern, drew his sword, and killed the man. Friends with influence obtained his freedom from jail, but he died mysteriously while awaiting trial, rumored by many to have taken his own life.

Superintendent Herbert and the miners kept things going for a time, but financial difficulties soon befell them. Bringing fresh money, a Colonel Charles Lynch bought into the operations in 1767 and stayed involved in the affairs of the mines for the next 20 years. Although his identity is still disputed, this is apparently the man whose name gave rise to the infamous terms "lynching" and "lynch law." In the 1780s Lynch led a group of friends and supporters who punished Tory sympathizers in southwestern Virginia during the American Revolution. Suspects received a brief sham trial in an unlawful court where the sentences involved beatings, loss of property, mandatory pledges of allegiance, and sometimes forced service in the military.

To this day, outlaw punishment, including murder, of vigilante victims bears the name of a man once intimately involved in running the Wythe County lead mines.

Despite the many struggles besetting the lead works in the years following Colonel Chiswell's death, it somehow managed to stay in business. As war with Great Britain threatened in 1775, a group of colonists met at the little settlement called the Lead Mines and drew up the Fincastle Resolutions. This document, sent to the Virginia delegate to the Continental Congress, pledged support for the Congress's resistance to the "intolerable acts" of the British crown. This was among the earliest declarations in the colonies that expressly vowed a fight to the death to preserve political freedoms and, some later claimed, provided part of the basis for the July 4, 1776, Declaration of Independence.

When the Revolutionary War broke out in April 1775, lead was not plentiful in the colonies and much of the available stores had been imported from France. Thus the Wythe County lead works immediately rose to the fore as the main up-and-running supplier of ammunition for the muskets of the colonial armies. Throughout the conflict British loyalists in the region posed a real danger (as did the Indians that sometimes cooperated with them), yet they never succeeded in capturing or disabling the mines. With independence won at Yorktown in 1781, tranquility returned to Southwest Virginia and the lead installations.

By this time, the Wythe County lead operations had expanded significantly. A 1782 report on the status of the lead mines noted that "...the furnace, the stamping [crushing] mill, and the grist mill were in repair, and the furnace house was 'very indifferent.' The dwelling houses and Negro houses were in 'tollable' order" [101]. A variety of equipment functioned to dig and process the ore. "At the furnace and washhouse there were paddles and molds for casting the bars of lead, wash tubs, sifters, shovels, shot molds and cards for dropping small shot through" [101]. The mines themselves housed "...numerous tools including axes, hammers, shovels, chisels, and mallets" [101]. Crops and animals grew on site to feed the mining community, which included 31 black families, the males evidently working as hired slaves.

In 1789 a major change in ownership took place at the lead mines. Two brothers, Moses and Stephen Austin, formed a partnership and secured a ten year lease on the facilities. They rebuilt the somewhat dilapidated processing plants and storage buildings, and employed additional personnel to help manage the plant. The industrious Austins ramped up the work force to 60 men and production jumped to a ton and a half of lead per day. Around this time the growing community became known as Austinville, and there Moses' son, Stephen F. Austin, was born in 1793. Five years later, little Stephen left with his family, some slaves, and mine laborers for Missouri. Moses had visited and bought a lead mine in this part of the Tri-State District the year before and now he sought his fortune in the West. Eventually his son would migrate farther south and found the Republic of Texas in 1836.

Moses' brother Stephen continued to make lead at Austinville for a while before mounting debts in the early 1800s cost him his interest in the mines. Ownership changed hands over the next few years, then Thomas Jackson and two other men bought the operations for $19,000 at the state auction in 1806 as noted earlier. Jackson was another English immigrant with skills in "...discovering, raising,

Fig. 6.5 Thomas Jackson's Shot Tower located along the banks of the New River near the Wythe County lead mines. This structure dates back to the first decade of the nineteenth century or possibly even earlier

washing, and smelting the lead mineral" [102]. He served as one of the experts brought in earlier by the Austin brothers and possessed extraordinary drive and ambition that revitalized the works at Austinville. In their time as mine owners, Jackson and his partners had use of the Shot Tower (Fig. 6.5). This still-standing magnificent dolomite block structure is on a high bluff overlooking the New River about three miles distant from the lead mines. It is one of the few shot towers remaining in the United States and presently is a Virginia state park.

Jackson has been credited with building the Tower sometime in the first decade of the 1800s, but the historical record is far from clear. A Virginia newspaper in 1791 mentioned a "Shot Factory," a name given to such edifices in those days, in Southwest Virginia. William Herbert, whom Chiswell had brought over from England in the 1760s, knew shot tower technology very well, as his employer in England had one constructed at Bristol. Herbert may have built one along the river well before the nineteenth century. Moses Austin referred to the manufacture of shot in his records, noting that the different climate in America necessitated a significantly different tower height from those in England. At any rate, if Thomas Jackson did not erect the original Shot Tower, he certainly must have rebuilt or improved an already existing one.

Shot towers were tall structures designed to generate small lead pellets in a range of sizes used primarily for hunting. The tower near the lead mines is 75 foot top to ground with a lower shaft dug another 75 feet down to the level of the river. The smelters at the digs sent over wagonloads of lead bars, which oxen, pulling block and tackle, hoisted to the top of the tower where a furnace re-melted the metal. Workmen poured the hot liquid lead through sieves of various sizes depending on the shot size desired. The lead droplets formed nearly perfect little spheres falling through the air during the 150 foot drop. The metal beads plunged into a quenching kettle filled with water at the bottom of the shaft; from there more laborers (probably slaves) pushed the cooled shot through an access tunnel to the river. Different sizes were sorted out, then carried away by wagons or barged on down the river to markets throughout the Southern states.

Once they acquired ownership, Jackson and his partners feuded almost continuously for the next two decades over the mines management and profit sharing. The War of 1812 renewed demand for lead by the American forces, and Jackson worked hard for his adopted country. After peace came, he engaged more of his native countrymen to assist with the lead works. Despite the difficulties with his partners, the business increased considerably. Upon Jackson's death in 1824, arguments and turmoil went on among the fractious mine owners, many of them descendants of Jackson and his original co-owners.

In 1838, a new operating entity came into existence, the Wythe Lead Mines Company. Over the next ten years, the mines prospered as the less contentious management made improvements and procured fresh markets, particularly in the North. When the contract ended in 1848, the operators changed the firm's name to the Wythe Union Lead Mine Company and had another ten-year agreement drawn up. Renewed discord among the owners followed, with one partner ending up constructing his own smelting furnaces, ore washer, and a water wheel to drive the machinery. At one point in the on-going litigation, some of the workers bore witness to their "lead colic," surely a manifestation of the metal's toxicity.

On March 8, 1860, under thickening war clouds, the Union Lead Mine Company was incorporated for the purpose of mining and manufacturing lead and shot. Throughout the Civil War, the Confederacy's "...principal, and most reliable, source of lead..." [237] would be the ironically-named Union mines in Wythe County, Virginia.

Notes

The story of Thomas Jackson's trek across early nineteenth century Virginia was told to me in 1995 by Mr. Pete Spraker, longtime guide and historian at the Shot Tower State Park. The other information about Jackson is from Kegley [103] and a brochure about the Shot Tower published by the Virginia Department of Conservation and Recreation [5]. Coggins [32] provided much information on the development of lead bullets, especially the Minie ball, and Civil War firearms. For "A Brief History of Lead" I relied on Jensen and Bateman [95], Craig et al. [36], and internet sources, Corrosion Doctors [34] and Wikipedia [264]. The section concerning the origin and evolution of "Firearms and Bullets" is based on information found in Moore [157], Wallace [249], and Wikipedia [262, 265] internet sites on the musket and bullet. The geology near the beginning of "The Wythe County Lead Mines" discussion is from

Pfeil and Read [176] and Sweet et al. [230]. Within that same section, the early history of the Wythe County lead mines is grounded most heavily in Kegley [103]. Her Chapter 14 on the county's historic mining industries is by far the best single source for this topic. Information in this section also comes from Watson [250], Vandiver [242], Donnelly [47], Austin [12], and Robertson [193].

Chapter 7
The Lead Mines Under Attack

"The Yankees are coming at dawn"

One hundred and fifty years later, the people of Wythe County, Virginia, still talk about the fabled ride of Molly Tynes that saved Wytheville and the lead mines close by. Molly, a cultured young woman who had attended college, was 26 years old that summer of 1863. She had come back home to help her father, Samuel Tynes, take care of her invalid mother on the family farm in southwestern Virginia. That same summer, Federal planners had determined to dispatch a raid deep into the very same country. The objectives were to attack Wytheville, eliminate the lead operations a few miles south of town at Austinvlle and the salt works at Saltville, and sever the Virginia and Tennessee Railroad.

On July 17 near midday, Samuel Tynes got word that a sizeable Union detachment, perhaps over a 1,000 men, had camped the night before on another farm about a mile away. Not long after this, two or three soldiers in blue rode onto the Tynes place and Samuel learned from the talkative troopers their numbers and intentions. Wytheville and the people along the way had to be warned, but how? The old gentleman did not want to leave his daughter and ailing wife unprotected; however, Molly's pleas at length convinced him to let her go in his stead. It would be a difficult task, for between the Tynes farm and Wytheville lay 41 miles of some of the most precipitous terrain in the western Appalachians, including four tall, thickly timbered mountains.

As fast as she could, Molly saddled up her mare Fashion and galloped away, stopping at every cabin she saw and crying "...the Yankees are coming at dawn" [40]. She rode on into the night, crossing Walker Mountain, the last and highest of the sandstone-crested barriers, in the pitch blackness of the pre-dawn hours. With torn clothes and bleeding from several cuts, Molly and an exhausted Fashion pounded onto the streets of Wytheville as the rising sun crimsoned the far horizon. When the Union riders arrived late in the afternoon, a small group of defenders

© Springer International Publishing Switzerland 2015
R.C. Whisonant, *Arming the Confederacy*, DOI 10.1007/978-3-319-14508-2_7

greeted them with stiff resistance. The Federal commander fell early in the ensuing Battle of Wytheville and his unit withdrew back into West Virginia that evening. The Northern invaders would return before long and in greater strength, but this time the doughty Miss Tynes had saved the town, the lead works, and a railroad upon which the Rebellion's ability to keep fighting mightily depended.

Significance of the Wythe County Lead Mines

The outbreak of war found the Confederacy woefully short of lead. From the start, the Wythe County mines at Austinville constituted the South's largest and most dependable source, but alone could not generate all of the lead demanded by its immense armed forces. Though importation initially yielded a substantial amount of lead, that conduit diminished as the conflict went on. Running the blockade was always a precarious business at best, and the dangers only increased as the cordon of Union ships tightened. Nevertheless, an early January 1865 report noted that nearly two million pounds of the metal had been imported by then, about 20 per cent of the Rebel consumption for the entire conflagration. By the end of hostilities, Austinville and importation had provided the great majority of the Southern lead.

During the war, lead became increasingly scarce and desperate Confederate administrators never stopped trying to develop additional means of procuring the metal. The government asked citizens for contributions of common household items containing lead, such as pipes, roofing materials, window weights, and eating utensils. To help alleviate the ongoing shortages, no possible source of lead for recycling could be overlooked. In 1863, the city of Mobile ripped up unused water mains and shipped them off to the arsenals. On occasion, officers directed soldiers back onto battlefields after the fighting ceased to scavenge for bullets. Spent ammunition could quickly be recast at the big armaments factories into fresh rounds.

A few other lead mines existed within the borders of the Confederacy, for example in North Carolina, Tennessee, Arkansas, and Texas, but these were modest workings active only sporadically during the war. Keeping the Western Theater forces supplied with lead posed an especially daunting problem. As early as 1862, a supply official in the region observed that "...the deficiency of lead does not permit more rapid fabrication of cartridges" [238]. The next year Ordnance Department Chief Josiah Gorgas wrote that "The question of the lead supply is nearly if not altogether as vital as that of niter..." [239], describing the shortfall as a "crisis." Gorgas noted later in 1863 that "...the Richmond laboratory, which made most of Lee's ammunition, could be supplied from the Wytheville mines, but... very little would be left over for shipment to other places" [240].

Once the war started, it did not take Confederate officials long to realize that the Wythe County mines would be of paramount importance as the principal domestic source of lead. The location of these installations in a remote mountainous sector made them relatively secure, plus the area had a modern railroad available to transport the lead pigs. Soon after fighting began, Richmond administrators "...demanded that the management either work the mines to their utmost capacity or surrender them for operation by the government" [46]. Directors of the Union Mine Company chose the former and round-the-clock activity commenced. The Ordnance Department monitored activity and did what it could to help by providing supplies and labor, even finding extra slave workers when conscription drew down manpower later. Such items as blasting powder, packing kegs, and tallow (for grease and candles to light the underground tunnels) arrived frequently at the works from the government bureau.

The Austinville facilities did not make the actual bullets. Rather, the workmen dug out the ore, processed it in the smelters, cast the molten lead into ingots, and hauled the lead bars by wagons to Max Meadows, a railroad shipping point about ten miles north of the mines. From there, the Virginia and Tennessee carried most of the bulk lead to factories in Richmond and Petersburg to be made into ammunition; some also went south to Knoxville and Chattanooga in Tennessee. Lead shot continued to be manufactured at the old Shot Tower not far from the excavations and a small number of bullets may have been cast by hand there as well. At times, the mines contributed more than refined lead, for example shipping slag—lead-rich waste from the smelting process—to be further processed at the Petersburg Lead Works in 1862.

Surviving output records are incomplete yet indicate that from May 1, 1861, to December 17, 1864, the Austinville mines provided at least 3.3 million pounds of lead. In addition, another 870,000 pounds went to various Southern states between February 1860 and May 1861, most of which probably ended up being utilized during the hostilities. In all, the Wythe County works produced roughly one-third to perhaps 40 per cent of the estimated ten million pounds of lead used by the South in the Civil War. Had the Federal high command recognized the true significance of the mines when open warfare got underway, earlier and stronger attempts to eradicate them might have been ordered. As it was, the North did try to get at Wytheville and Austinville from time to time, albeit usually as part of larger strikes against the seemingly more valuable salt works and railroad. In 1864 Northern strategists at last began to appreciate the Wythe County mines as the most worthwhile target in southwestern Virginia. In December of that year, less than four months before the end of the war, the Union mounted an attack in real strength on the lead mines. But by then, the metal turned out by those operations had done incalculable damage to its soldiers.

The Battle of Wytheville

Wythe County was created in 1790 and named for George Wythe, a signer of the Declaration of Independence and an important figure in early American jurisprudence. By the 1860s, the bustling community of Wytheville, the county seat, boasted a population of about 1,800 people. The town lies in the Great Valley guarded over by Lick Mountain, a long quartzite ridge abruptly rising over 3,000 ft just south of town.

Ten miles farther south on the New River were the lead mines, 30 miles to the northeast stood Central Depot and the New River railroad bridge, and 40 miles to the southwest lay Saltville. Wytheville had a depot on the Virginia and Tennessee line, and along the tracks west of town the "High Bridge" over Reed Creek made an inviting target, second only to the Long Bridge at Central Depot, for Federal raids intended to cut the rail line. A crucial wartime industry in Wytheville, Barrett's Foundry, served as one of the few places in the South able to make percussion caps for infantry rifles. Several roads and turnpikes intersected at Wytheville, among them some from West Virginia that Northern incursions could and did utilize as invasion routes to get at the mineral industries and railroad. All in all, the central location of Wytheville in such a convenient cluster of objectives for Union strategists made it extremely likely that sooner or later the soldiers in blue would arrive.

The first of those assaults came in 1863. The initial two years of the conflict had been relatively quiet, even prosperous times in Southwest Virginia. The war industries boomed and the fighting seemed far away as the main armies locked in bloody battle at places like Manassas, Shiloh, Fredericksburg, and Chancellorsville. Then, over the first four days of July 1863, two disasters befell the Confederacy—Lee's defeat at Gettysburg and Grant's capture of Vicksburg. The tide of combat had turned decisively in favor of the North. Now, more resources could be directed to secondary arenas like southwestern Virginia and its lead, salt, and railroad operations. For months, Union officers at the large army base in Charleston, West Virginia, had been marshaling men and equipment in anticipation of just such an expedition against those objectives. The time to launch that campaign was at hand.

On July 13 Colonel John Toland left Charleston with about 1,000 cavalry and mounted infantry under orders to strike at Saltville, Austinville, and the Virginia and Tennessee (Fig. 7.1). Toland decided to move on the salt works first, entering Virginia and then following a narrow valley toward the salines. On July 17 the Union colonel and his horsemen met and overpowered a small Confederate outpost blocking their advance, taking prisoner all save one of the enemy soldiers. Fearing that this lone Southerner would forewarn Saltville and its several hundred defenders, Toland switched his mission to an assault on Wytheville and the military stores and factories there. He also intended to sever the railroad by destroying the High Bridge west of town, and then proceed to Austinville to take down the lead operations.

Turning his men to the new goal, the Union colonel had them up early the next day and underway by 3 a.m. Late in the afternoon on July 18, Toland and his troopers reached the outskirts of Wytheville. The town had been warned, thanks perhaps to Molly Tynes, and hastily assembled about 120 local civilians and a few soldiers passing through to repel the invaders. By now General Sam Jones, commander of the Department of Southwest Virginia headquartered in Dublin, was aware of the threat. Rounding up every man he could find, he sent 130 regulars under Major Thomas Bowyer by train down to Wytheville where they joined the gathered defenders.

Around 7 p.m., the Union men spurred their mounts and started trotting into the downtown area of Wytheville, then paused for a moment before Toland ordered a charge. A volley of Confederate musketry ripped into his riders, knocking many

Fig. 7.1 Union Colonel John Toland. Image from MOLLUS-MASS Collection, U. S. Army Military History Institute. Toland led the July 1863 attack on the lead mines, but was killed in action at Wytheville

from their saddles and slowing the onslaught. Amid persistent fire from the windows, doors, and front yard fences of houses lining the street, the Federals reformed and tried again to advance. When this effort failed, Toland ordered them to dismount and attack on foot, making his four to one advantage in numbers that much more telling. His charges did so and as the colonel watched them press forward, a bullet hit him squarely in the chest, killing him instantly. The Union forces kept going, and the Confederates, taking mounting casualties, steadily fell back from the buildings. Major Bowyer saw that the day was lost and gave the order to retreat, leaving the town in the hands of the United States army.

The Battle of Wytheville lasted less than an hour, yet in that time, the Northern leader and other officers and regulars had fallen. Although victorious, the Union command had been decimated, leaving the ranking surviving officer unsure of what to do next. After a short while, he simply gathered his troops and turned back to West Virginia. Left behind were about ten killed and perhaps as many as three dozen wounded soldiers; Southern casualty counts were similar.

The results of all this proved negligible. Before departing, the men in blue set parts of Wytheville aflame and damaged portions of railroad tracks that later took about an hour to repair. The High Bridge remained untouched and the lead mines never attacked. The mines home guard, consisting of two companies of workers, had been called out to help defend Wytheville. They arrived too late to be of any consequence and went back to the mines.

As for the defeated Confederate soldiers brought to the fight by rail from Dublin, one last indignity awaited. Major Bowyer had firmly instructed the engineer to keep the train that got his detachment to Wytheville waiting at the depot should they need to withdraw speedily. Nonetheless, when Bowyer and his retiring warriors returned to the station, they found the locomotive already departed. The panicked driver had thrown the wheels into reverse and was backing the train to the safety of Dublin. It was indeed a disgusted and dispirited group of Southern troops, who first having lost the firefight for Wytheville, then had to walk the 25 miles or so back home to boot.

In Wytheville after the battle, the town women may have prevented a massacre. Enraged at what the Yankees had done, many of the male citizens talked of hanging some of the wounded bluecoats. The wives urged their men folk to be more merciful and no one ended up on the gallows. Colonel William Powell, the top Union officer left behind, became the focus of the proposed rough justice. Three local women, in particular Mrs. Susan Spiller who had two houses she owned burned by the intruders, hid him until the danger passed. After peace came, Colonel Powell traveled back to Wytheville to visit Mrs. Spiller, and he remembered her for years with Christmas gifts. A few days after the battle, Mrs. Toland and another Union widow journeyed to Wytheville to claim their husband's bodies. The town ladies received the grieving women with warmth and sympathy, a touching moment of kindness in a war filled with violence and cruelty.

The Battle of the Cove

No real menace to the lead mines reappeared until almost a year later. In March 1864 General Ulysses S. Grant took control of all the United States armies and by May had forces on the march throughout Virginia. His grand strategy aimed to assail the Army of Northern Virginia from a number of fronts. As part of this renewed offensive, Grant ordered General George Crook's Kanawha Division, based in Charleston, West Virginia, to strike deeply into Southwest Virginia. The prime goals of the mission were to destroy the railroad bridge over the New River near Central Depot and, if possible, devastate the salt works at Saltville. Should this be accomplished, Crook had instructions to link up with another Union army driving south in the Shenandoah Valley, shutting off any retreat by Lee's hard-pressed army in eastern Virginia into the western mountains.

Crook's army of about 6,200 men plus artillery departed from Charleston on April 29. Along with Crook rode a contingent of 2,500 mounted troops under General William Averell (Fig. 7.2). Colonel William Powell, part of the Toland raid a year earlier when the Wytheville women saved his life, led one of Averell's regiments. Within a few days, Southern informants passed on information to the Confederate high command that Crook's division was on the move. General John Hunt Morgan was promptly directed to block this latest Union incursion (Fig. 7.3).

Morgan had earned a reputation as a good field officer and very aggressive cavalryman. His 1862 raid into Kentucky and an 1863 slashing attack across the Ohio River into Indiana and Ohio brought forth much praise in the South and consternation

Fig. 7.2 Union General
William Averell. Photograph
courtesy of the Library of
Congress. Averell
commanded the May 1864
assault on the lead mines that
failed to reach the vital
mineral works

Fig. 7.3 Confederate
General John Hunt Morgan.
Photograph courtesy of the
Library of Congress.
Morgan's rapid response
to the May 1864 Union
incursion saved the lead
mines for another seven
months

in the North. Morgan took over leadership of the Department of Southwest Virginia
in March 1864 and was in Saltville that May when he received the reports of a major
enemy concentration headed in his direction.

As his columns neared Virginia on May 5, Crook split off Averell to assault
Saltville while he kept moving on to the New River Bridge with the main army.
Averell was a seasoned officer who had led a daring raid against the Virginia and
Tennessee Railroad earlier in December 1863. Moving deeper into the mountains
on his present mission, Averell received intelligence that Morgan was defending the
salt works with about 4,000 men and artillery, outnumbering his unit considerably.
The strength of the Saltville defenses was greatly exaggerated, but the misinformation

caused Averell to shift his target from the salt operations to Wytheville and the lead mines beyond. Approaching the town, the Union commander's troopers camped on the night of May 9 in the Cove Mountain area a few miles north of Wytheville.

Morgan, now seeing that Averell planned to attack Wytheville and the lead mines, raced north and reached Wytheville before his opponent. Finding some Southern troops already there, the Confederate general sent them to guard the gap passing through Cove Mountain. Around 3 p.m. next day, May 10, Averell encountered Morgan's outpost, and the soldiers and citizens in Wytheville heard the sounds of battle once again. Morgan rushed his command to the shooting and, arriving in strength, began to outflank the Union horsemen. Averell saw that he could not push his way through the mountain pass, and withdrew a short distance to set up a new position. During this action, he barely escaped death when a musket ball grazed his scalp. The determined Federal leader stayed in the fight, however, shouting directions as blood soaked the handkerchief binding his wound. Hard beset by Morgan, Averell pulled back another several hundred yards, this time halting at an old farm house where his men threw up defensive lines as best they could. The ever aggressive Southern commander kept pressing his foe; at one point Morgan manned one of his own artillery pieces with effective results. But night was falling, and the hope of a complete Confederate victory faded with the daylight. Again the order for Federal retreat sounded, and the Northerners slipped away in the darkness, heading northeast to rejoin Crook and the main body.

The Battle of the Cove was over, resulting in casualties totaling about 120 on the Union side and 40 for their opponents. Once more Wytheville and the lead works had escaped destruction. Morgan wrote afterwards "If I had been one hour later, this place and the lead mines would have been lost" [51]. Averell had a different version of the engagement, noting in an after-action report to Crook that he fought the Confederates for four hours, "…inflicting some loss and capturing a few prisoners…" and that "… the enemy retired after dark" [52].

Interpretations of the tactical outcome of the fighting at the Cove may have varied, yet one hugely important strategic result remained indisputable. Union failure in this seemingly inconsequential mountain battle meant that Wythe County lead, by then practically the South's sole supply, would go on providing Confederate armies with ammunition to keep them in the field. For another agonizing seven months, when Federal raiders finally overran the mines, Austinville lead production went on unabated.

Stoneman's Raids

By December 1864 a rapidly weakening Confederacy, shrunken by the loss of territory and faced with a sharply declining pool of men available or willing to fight, struggled to survive. Grant relentlessly increased pressure on Lee and the Army of Northern Virginia who were trying to hold Richmond and Petersburg. In southwestern Virginia, Union scouting parties sometimes roamed at will. The citizens of the region, besides facing Union invaders, were frequently terrorized by outlaw bands

Fig. 7.4 Union General
George Stoneman.
Photograph courtesy of the
Library of Congress.
Stoneman's raids in
December 1864 and April
1865 seriously damaged the
lead mines operations, but did
not destroy them completely

of bushwhackers, murderers, and deserters from the Confederate army. Still, regardless of the late stage of the war and the few regular servicemen left in Southwest Virginia, the three primary targets of Federal destruction—Austinville, Saltville, and the Virginia and Tennessee Railroad—stood intact and operational.

General George Stoneman, an ambitious Union cavalry commander, resolved to change all this (Fig. 7.4). At this point in the war, his career was in tatters and his own Secretary of War, Edwin M. Stanton, had damned him as "…one of the most worthless officers in the service" [55]. Stoneman had indeed performed ineptly at Chancellorsville in 1863 and his inexcusable capture with 700 of his men in the Atlanta campaign in summer 1864 only made matters worse. Exchanged a few weeks later and now in command in East Tennessee, the disgraced general saw a successful raid against the mineral works and railroad in southwestern Virginia as the key to salvaging his reputation.

Stoneman left Knoxville, Tennessee, on December 10 with 6,000 mounted troopers plus artillery. The Federal units proceeded northeast along the railroad, entering Virginia at Bristol. Following the rail line up the Great Valley corridor, Stoneman drove Confederate defenders before him, tearing up Virginia and Tennessee tracks and burning trestles, rolling stock, and depots. At Marion the Union expedition overwhelmed a thin opposing force, chasing it to Wytheville. The Southerners abandoned the town and Stoneman's men moved in on December 16, seizing stores of munitions and setting some buildings ablaze. Next day the Northern commander sent two regiments of troopers to unleash destruction on the lead mines.

Stoneman's raiders met no attempt to defend the lead mines on December 17; the few Confederates assigned that task decided instead to retreat at their approach. The Union riders crossed the frigid New River, the biggest obstacle between them and the mines, without incident and set about demolishing the lead works. In only two hours, the mine offices, storehouses, stables, crushing machine, bellows, furnaces, sawmill, and gristmill went up in flames. The wreckage would have been much worse but the mines superintendent eventually raised the white flag, avoiding loss of all the property. Retiring back south in the Great Valley, Stoneman turned toward Saltville where he broke through much reduced defenses on December 20. Next day saw wanton demolition visited upon the salt operations, after which the Federals returned to Knoxville, reaching that city on December 29. Behind Stoneman lay ruined railroad tracks, engines, cars, depots, and bridges. Even though the lead works had been seriously damaged, repair crews got busy immediately and on March 22, 1865, the mines went back on line.

By this time, all of the Confederacy's accumulated reserves of lead had been used up and the armies in the Eastern Theater relied completely on the day-to-day output of the Wythe County mines. The Union knew that the Austinville installations were back up and running, and more necessary than ever to the Southern cause; thus orders came down for a new thrust to obliterate them completely. The task fell once again to Stoneman, who then had his soldiers in western North Carolina positioned to cut off any reinforcements for the beleaguered Lee that might be coming through that area. The Confederates, meanwhile, had anticipated another attack on the Austinville lead works in spring of 1865 and maintained a small force in the vicinity should they undergo another assault.

The final strike on the Wythe County mines got underway in early April when columns from Stoneman's command again advanced into southwestern Virginia. A brigade of 500 men under Colonel John Miller headed for Wytheville with instructions to sever the Virginia and Tennessee Railroad by wrecking two bridges near the town. Any military supplies in Wytheville were to be destroyed, then the lead mines leveled once and for all. Colonel Miller's troopers, fortuitously finding one of the mine foremen in their path, convinced him to guide them to the New River ford. Crossing over, they reached the town on April 5 and occupied it with no resistance.

In the meantime, the Confederates, having learned of the impending attack, dispatched an undersized brigade by train to Wytheville to turn aside the threat. Reaching the town, the Rebels chose to attack and drove away their foes in a brief skirmish. Before departing, Miller's horsemen devastated a commissary along with quartermaster and ordnance stores that included ammunition and 10,000 pounds of gunpowder. Next day, the retreating Northerners knocked out the High Bridge over Reed Creek on the rail line, then rode to the lead mines. With no guard of any kind to deter them, they laid waste to the partially rebuilt production facilities and the repair materials gathered to keep them going.

By then, none of this really mattered. Two days later on April 9 Lee surrendered, ending any further need for Southwest Virginia's minerals and railroad to sustain the collapsed Rebellion. Still and all, the Wythe County lead mines had made a fundamental difference in the Southern effort to keep its armies in the fight. After hostilities ceased, Confederate Ordnance officer Colonel William Broun summed up the contribution of the Austinville lead:

Our lead was obtained chiefly, and in the last years of the war entirely, from the lead mines at Wytheville Virginia. The mines were worked night and day, and the lead converted into bullets as fast as received. The old regulation shrapnel shells were filled with leaden balls and sulphur. The Confederacy had neither lead nor sulphur to spare, and used instead small iron balls and filled with asphalt. [192]

Despite all the intensive efforts to find more lead, no other domestic source ever approached the scale or importance of the operations along the New River deep in the mountains of Southwest Virginia.

Notes

The story of Molly Tynes' ride to save the lead mines is from Daniel [41], a very traditional retelling of this thrilling tale. Kegley [104] refers to the feat as "encrusted legend" (but what stories of American heroes and heroines aren't?) and details what probably really happened. The "Significance of the Wythe County Lead Mines" information comes from Vandiver [242], Donnelly [47], and Robertson [193]. The remainder of this chapter concerns the Civil War fights involving the lead works, and Kegley [103] contributed to the accounts of all these battles. Again, her thoroughly researched work [103, 104] on the entire history of Colonel Chiswell's mines (and Wythe County) are required reading for anyone interested in the subject. Concerning the specific military actions, I obtained much material from Johnson [99] for Toland's 1863 attack ("The Battle of Wytheville") and Emerson [53, 54] for the 1864 "Battle of the Cove" and "Stoneman's Raids" near the end of the war, respectively. The personal information about Stoneman is from Fordney [64] and Evans [56]. Donnelly [47] and Evans [56] also have much information on the raids late in the conflict. The books by Walker [246], Johnson [98], McManus [144], and Marvel [131] gave additional information on the Civil War military events in Wythe County.

Chapter 8
The Saltville Salt Works

"Salt is eminently contraband"

It is now the morning of December 21st, 1864. We were sitting round a blazing hickory fire chattering as merrily as Confederate women could in that disastrous year, trying to banish sad thoughts of the sufferings of our poor soldiers in those dark days – for a group of little boys were asking endless questions about Santa Claus. [223]

So begins Ellen Brown Stuart's eyewitness account of a Christmas season in Saltville when it fell under attack by a United States army. Ellen Stuart was the 26-year-old wife of William A. Stuart, one of the principal owners of the South's largest salt-making operations. William, the eldest brother of Confederate General Jeb Stuart, had brought Jeb's widow Flora and her two children to Saltville following the iconic cavalry hero's death at the Battle of Yellow Tavern earlier that year. Flora would spend the rest of the war and several years afterwards in the town teaching school and raising her family under William's care.

Saltville had been menaced by Northern invaders on several previous occasions and an important battle had erupted there three months before in early October. At that engagement, mounted Union troops from Kentucky under General Stephen Burbridge had been driven away, but the Federals returned in December with a stronger force commanded by General George Stoneman. This time the enemy horsemen faced significantly fewer Confederate numbers and had broken through, determined to reduce the salt-making facilities to ashes completely.

Ellen Stuart's narrative continues with the small lads wondering whether the old elf could elude the armies: "Little Jeb interpolated maybe Santa could shoot us some fireworks as there are no toys, books or candies. Ah! my boy you were such an unconscious prophet, but somehow they had faith in old Santa just as we grown up people had in God" [223]. A loud knock at the door ushered in news that cavalry under Generals Burbridge and Stoneman were rampaging through the countryside. Advised to leave the house, the Stuart family sought refuge with the household of George Palmer, a salt maker from New York who was now a partner with William

© Springer International Publishing Switzerland 2015   87
R.C. Whisonant, *Arming the Confederacy*, DOI 10.1007/978-3-319-14508-2_8

Stuart in the business. By 9 p.m. that evening, the sounds of musketry and cannon had ceased, and three Confederate officers, one of them Ellen's husband William, stopped by to announce that Saltville was being evacuated. The few remaining defenders had abandoned the salt works to their fate. Ellen writes:

> The night was dark as Erebus and bitterly cold, the ground covered with sleet. We could but think of Valley Forge. The Federals soon entered the place, advancing cautiously however. Toward midnight a grand illumination succeeded as the furnaces and machinery for pumping the salt water were blazing, then the depot was blown up, so Jeb had 'his fireworks' before Christmas, as the other boys said. [224]

Next day Mrs. Stuart got a note to General Burbridge, who was occupying her house, asking for a "…pair of mules or horses… that I might return with my little children to my home" [224]. She received a courteous reply from the Union officer that he would indeed send an ambulance to transport her family, but before that could happen, rumors that Confederates were coming in strength got the raiders "… all put out for Kentucky in a hurry" [225].

Despite the hearsay, the Union raiders had not fled in panic. They remained in the valley another full day, torching the salt sheds, smashing the big iron kettles, and throwing pieces of debris, including torn up railroad tracks, down the brine wells before departing. Although Northern victory in the December 1864 Battle of Saltville inflicted much damage on the salt works, the devastation of this enormous production complex was far from complete. Within a few weeks, the strategic mineral would be pouring forth from the wells and sheds again, sustaining the armies of the Confederacy while the last flames of hope for separate nationhood sputtered out.

Of all Virginia's mineral contributions, perhaps none held more importance to both the general population and the military forces of the Confederacy than salt. Salt is essential in the human diet and in the Civil War, every soldier's ration included it. Furthermore, salt is necessary for livestock; a lack of salt apparently caused the hoof and mouth disease that appeared among the cavalry horses of Lee's army in 1862. In Civil War times, salt served as the primary means of preserving meat. Additional uses included packing certain foodstuffs, particularly eggs and cheese, and preserving hides for making leather. Armies of that day demanded tremendous amounts of leather for soldiers' shoes and accessories, and for horses' bridles and saddles. Beyond these direct military needs, the mineral found use in various chemicals and medications required by both citizenry and armed services. Early in the conflict in 1862 Union General William T. Sherman branded salt as an utmost strategic resource: "Salt is eminently contraband, because of its use in curing meats, without which armies cannot be subsisted" [113].

By the mid-nineteenth century salt production typically involved three methods: extraction from saline water wells (the most common), boiling down sea water or water from inland salt lakes, and mining deposits of rock salt. At the outbreak of the Civil War, the Southern states had five principal salt operations available, these being the "Licks" on the Great Kanawha River near Charleston, Virginia (after June 1863, West Virginia); the Goose Creek Salt Works near Manchester, Kentucky; the wells in the counties of southwestern Alabama; Avery Island in southern Louisiana;

and, above all, the wells in southwestern Virginia at Saltville. In addition, various places scattered along the Confederate sea coast manufactured salt and a considerable industry of this type sprouted up in Florida as the war continued.

The South lost the Goose Creek works almost immediately after the fighting began, as well as the facilities on the Kanawha. The fall of Vicksburg on July 4, 1863, denied all of the extensive Louisiana sources to the eastern Confederacy. Thus, by mid-summer 1863, even though the Alabama wells still operated in the Gulf Coast area, the Stuart, Buchanan, and Co. salt works in Smyth County, Virginia, had to supply the states east of the Mississippi almost alone. The presence of these crucial salt operations, together with the lead mines in southern Wythe County and the Virginia and Tennessee Railroad, attracted intense interest from the North throughout the years of hostilities.

An Ancient Mineral Resource

Salt is one of the most abundant mineral resources in the world, forming anywhere briny waters undergo significant evaporation. Its use by humans is long preceded by the need of mammals to have sodium and chlorine, the two elements that make up salt, in their diets. Indeed, the prodigious amounts of meat eaten by the early peoples probably gave them enough of this essential nutrient, yet that presented a problem: how to preserve the animal flesh to ensure that an adequate amount would be available between hunts. That and the founding of cereal- and grain-based farming societies that used salt as a food additive rendered the mineral a valuable commodity and perhaps a currency before any record of such use.

Sodium and chlorine make up 87 per cent of all the dissolved elements in sea water. Hence, obtaining salt from marine brines by evaporating away the water is relatively easy and has been done in coastal quarters for thousands of years. Surface deposits of halite, the mineral name for common salt, also naturally abound, particularly around the margins of inland salt lakes and seashores in the warm, dry climes of the Middle East, Africa, and the Orient, all cradles of civilization.

Historians speculate that around 8000 BCE the Chinese gathered natural salt crystals formed on the surface of a brine lake in the northern part of their country. By 2000 BCE salt was used in China to preserve fish, and salted fish and birds are known still earlier from Egyptian tombs. Later, both of these cultures learned to place sea water in containers, evaporate the water, and harvest the solidified salt. When Rome embarked upon its ascendancy to political and military power, a brisk trade in salt and salt-cured foods already existed around the Mediterranean. The Romans manufactured their own salt by 640 BCE, collecting ocean water in large pools where the sun's rays caused the halite to crystallize. One of the earliest main Roman roads was the Via Salaria—the Salt Road. This highway brought salt taken from brine ponds on the Adriatic coast to Rome across the Italian peninsula. Along the Adriatic shoreline, marine waters are much more saline and therefore more productive than those from the Tyrrhenian Sea near Rome.

In the long history of Roman rule, salt remained a basic commodity for everyday life. Supplies came from many operations using different technologies scattered across the Empire. Most resided in the coastal regions where sea water evaporated in solar ponds generated the white crystals. In some of these works, laborers used fire to heat ocean water in clay pots, then broke them apart once the halite precipitated into a solid mass. Away from the shorelines, this technique could be used where brine springs flowed. In other places, miners dug ancient halite from the earth or scraped up recent deposits from inland salt pans.

The Roman government manipulated the price of salt from time to time, often taxing it to pay for wars, construction works, and maintenance projects. Citizens put it in their foods, as in bowls of fresh greens from which the name "salad" comes. Vintners added the mineral to wine, primarily to keep it fresh but for taste as well. Enormous amounts of salt went for meat preservation, especially fish and pork, and such staples were a principal component of commerce and trade in the Roman world.

After the Roman Empire fell in 476 CE, the Mediterranean basin endured as a major realm of salt making and merchants transported it far into central and northern Europe. During the Middle Ages, a substantial salt industry arose in northern Italy at Venice. The Venetians refined the single evaporation pond process by devising a series of pools that concentrated successively brinier waters until the halite crystals at last appeared. In 1281 the government of Venice started subsidizing the shipment of salt, boosting the City State into the most completely developed salt-based economy of all Mediterranean countries. At this same time, Genoa blossomed into a center of the salt trade and posed a serious threat to the domination of Venice. In the late fourteenth century, the two rivals fought a salt war in which Venice emerged victorious.

A century after the salt war in northern Italy, discoveries of other oceanic commercial routes besides the Mediterranean began the collapse of Genoa and Venice as maritime economic forces. In 1492 Columbus found a vast new world of resources across the Atlantic and 5 years later Vasco de Gama opened a sea lane around the southern tip of Africa to the Orient. In that same year of 1497 John Cabot, voyaging under the flag of England, came upon the rich codfish banks off the coast of North America. Salted fish, long a standard of global economic traffic, now assumed a value higher than ever.

Almost immediately the Atlantic seafaring nations of Europe set to competing for a share of the burgeoning salt cod industry. This put increased pressure on the production of salt, particularly sea salt from evaporated brines as this was thought better for preservation than mined rock salt. The Brittany peninsula on the French coast had the climate and geographic location for profitable solar pond salt making and such enterprises soon sprang up. Dutch, Danish, and British concerns became principal customers of French sea salt in the sixteenth century scramble for cod. By then the military powers of Europe had become highly dependent on salted meat, fish most of all, as strategically essential military supplies. Going to war now entailed laying aside huge stores of salt and all kinds of fish to feed the soldiers and sailors of the national armies and navies.

The mushrooming expansion of the European cod industry had spurred a pronounced upsurge in sea salt manufacture but salt also had a host of other uses as the Middle Ages drew to a close. It excelled as a preservative for a variety of foodstuffs and in the preparation of many more, notably cheese and butter. Salt played key roles in leather making, as a cleaning agent, and in the compounding of many medicines and chemicals. Since the days of Roman hegemony, underground deposits of halite had been mined in central and northern European countries such as Germany and Austria. Tradesmen in Poland had extracted salt from natural brine springs for millennia and in the mid-thirteenth century miners there opened excavations to obtain rock salt. Soaking vegetables in brines, specifically cucumbers and cabbage to produce pickles and sauerkraut, grew popular all across central Europe. Salt and salt-treated foods were vital commodities, moving about on the river systems and canals dug to move these goods with greater efficiency.

England ranked among the earliest leaders in finding and exploiting the prolific cod fishing grounds off the North American coast. As those endeavors escalated, so did the need for a reliable domestic source of salt. The ancient Britons had garnered salt from brine springs by boiling the waters in clay pots at Cheshire in northwestern England as early as 600 BCE. Upon their arrival in 43 CE the Romans refined this simple procedure by using the lead by-product from nearby silver mines to make containers. Craftsmen fashioned the lead into broad shallow pans in which heated brines turned out crystallized halite much more effectively than in the earthen pots. In the seventeenth century Cheshire was still the principal British salt-making locality. By then, the lead evaporators had grown enormously in size and the warming fuel was coal, not wood, the local timber stands having been nearly completely chopped down. The Cheshire salt was boated down the Mersey River to the port of Liverpool, making that city a premier exporter of salt.

Despite Cheshire's considerable production, England needed yet more salt, something that had troubled Queen Elizabeth in the late 1500s. Two centuries later the provision of adequate amounts of salt became an even greater concern. The British were then locked in an economic, political, and military struggle with France, and feeding every soldier and seaman required tremendous quantities of salt. By the late eighteenth century, England had won control of most of the marine fisheries off North America. Owing to this and a rapidly enlarging empire to defend, the demand for salt could only go up and the Americas would become a battleground for it.

Growth of the American Salt Industry

Native Americans made salt for thousands of years before Europeans arrived. Aztec, Incan, and Mayan societies produced it, as did many of the tribes in western North America, among them the Hopi, Zuni, and Navajo. Caribbean islanders were using sea salt when Spanish vessels first dropped anchor. Once ashore in the New World, Iberian settlers started up cattle herding and consequently required salt to feed the animals and cure the hides. The commencement of silver mining and processing demanded sizeable amounts of halite, leading to construction of nearby salt works. (The salt was thought to promote separation of the silver from its sulfide ore.) To the north the competition for the Grand Banks fishing industry drove British, French, and Dutch interests to seek fresh sources of salt, much of which eventually came from the Caribbean. Establishment of the 13 English colonies carried with it the need for more salt. Captain John Smith founded a salt works at Jamestown in 1607, the year of its settling. From that early beginning, Virginia persisted as a prime salt producer until the end of the Civil War.

A short time after Jamestown, salt manufacturing appeared elsewhere in the colonies besides Virginia, most notably in Massachusetts; still, the new Americans relied mostly on imported English salt for their supply. When the Revolutionary War erupted in 1775, salt stores diminished sharply as the British sources were cut off and the Royal Navy's blockade impeded importation from other countries. The Continental Congress responded swiftly to ease the shortage, passing in May 1777 an act granting a subsidy to all salt importers or producers. A year later Congress formed a committee "...to devise ways and means of supplying the United States with salt" [114]. A recommendation followed that the colonies promote salt production with financial assistance or by other considerations.

Setting up sea salt operations along the coast using solar energy offered the quickest and simplest way to increase manufacture, and many such endeavors quickly appeared. Cape Cod evolved into one of the earliest of the prominent salt-making centers. The region's cod fishing industry and unusually salty coastal waters provided a firm basis for the startup operations. Innovative salt makers soon learned to harness the strong coastal winds to power the pumps delivering seawater to the evaporators. Cape Cod generated salt throughout the war and well into the nineteenth century.

After the Revolutionary War, salt works got underway at a number of inland sites, most of them already discovered and utilized by the native peoples. In 1788 the Onondaga Indians signed a treaty that gave the state of New York ownership of lands with brine springs near modern Syracuse. Commercial salt entrepreneurs set to work immediately. By 1810 yield had expanded to around three million bushels (150 million pounds) a year, much more than anywhere else in the United States. At first, production stemmed mainly from heating the saline waters in big iron kettles set in stonework furnaces covered by open-sided sheds. The furnaces burned wood in immense quantities, soon exhausting the thickets of trees in the vicinity. Some owners brought in coal, but that proved expensive and resulted in many of the installations converting to solar evaporation techniques by 1820.

The presence of such an expansive salt industry in central New York provided a driving force for construction of the Erie Canal. When workmen at last completed the "ditch that salt built" in 1825, the Onondaga output soared. This led to the decline of the Cape Cod salt companies, although it took another four decades for them to dwindle away entirely. At Syracuse, a prospering town now known as the "Salt City", massive salt-making facilities went up, an expansion of operations that continued into Civil War times when production peaked at nine million bushels in 1862.

This dwarfed the maximum annual extraction of four million bushels at Saltville in 1864, the Confederacy's main source by far.

A few years after the hostilities concluded, an ex-Confederate officer journeyed to Syracuse to deliver a lecture. During the day he was shown the sights around the area, including the gigantic system of salt wells and production buildings. That evening he opened his presentation with this question: "Do you know why you northerners whipped us southerners?" [123]. Then the old soldier gave his answer: "Because you had salt."

Before the American Revolution, another major salt-making location emerged, this one in the extreme northwestern portion of Virginia, a region that would later become part of West Virginia. Along the banks of the Kanawha River near a wilderness settlement named Charleston bubbled large salt springs that had attracted Indians and game animals for many years. White pioneers made salt here beginning in the mid-eighteenth century, portions of which Daniel Boone carried west in his explorations. Commercial-scale brine processing started up in 1797 and by 1809 the Kanawha Licks, as the locale was generally known, were expanding rapidly. The need for salt in the War of 1812 added to a prosperity that kept going into the 1820s and 1830s. In those decades, Kanawha salt traveled on down the river and into the Ohio, turning towns like Louisville, Kentucky, and Cincinnati, Ohio, into primary salt-trading cities. The conjunction of Kanawha salt and Midwestern pork helped transform Cincinnati into the significant center of business and commerce it ultimately became.

Being in Virginia in the antebellum years permitted the use of slave labor at the Kanawha Licks, and they did much of the most difficult work associated with salt making. The worst such task was mining the plentiful local coal that fired the furnaces; blacks performed this dangerous job almost completely by themselves. Eventually the invention of steamboats allowed salt from other vendors to move up the Mississippi River and into the Ohio Valley, creating stiff competition for the Kanawha product. Even so, when North and South began shooting in 1861, these works stood second only to the huge Onondaga installations. Not long after, the operations in northwestern Virginia came under attack, falling to the Union and in time becoming part of the newly founded state of West Virginia.

Elsewhere in the South in the late eighteenth and early nineteenth centuries, salt makers erected works that would be invaluable when civil war broke out. Brine springs discovered at Avery Island in southern Louisiana in 1791 rose briefly to prominence in the War of 1812 when the high price of salt stepped up production in many places. Mostly idle in the decades before the Civil War, Avery Island got back on line in 1862 when the Confederate need for salt made operations profitable once again. That same year laborers digging a well encountered what proved to be a vast amount of pure, high quality rock salt at shallow depth. Numerous open pit halite quarries began and Avery Island provided considerable amounts of the strategic mineral until captured by the North in 1863.

Commercial salt interests started up at Goose Creek in southeastern Kentucky near Manchester in the 1790s. In 1861 five businesses extracted salt from these brine springs, but Northern forces ravaged them next year. Ironically, a Federal officer participating in the attack owned one of the plants demolished. In Alabama a few miles north of Mobile, salt making begun in the early 1800s decades later surged to significance as a noteworthy source, especially for provisioning the lower South. The briny spring waters, shallow and easily reached by wells, were heated in iron kettles until the white grains appeared. Weather limited output, however, to the months of April to December due to flooding from the Tombigbee River. Relatively safe in the swamps of south Alabama, these works manufactured salt until the war ended.

In spite of their continuous activity throughout the hostilities, the Alabama wells with their production restrictions and remoteness never came close to being the foremost supplier of Confederate salt. That responsibility fell to the salines clustered at a secluded village deep in the mountains of southwestern Virginia. Important from the very outset, for the last two years of the conflict Saltville churned out most of the salt used by the Southern military and civilians.

"Salt… of the purest quality" in Southwest Virginia

The town of Saltville, located in the northwestern part of Smyth County near the Washington County line, lies in a small valley within the Valley and Ridge province of the southern Appalachians (Fig. 8.1). Geology and human history here are closely connected, beginning with the arrival of Paleo-Indians in the Saltville Valley

Fig. 8.1 View of the Saltville Valley looking northwest. In the Civil War, a number of salt wells would have been seen on the valley floor. The Virginia and Tennessee Railroad and salt furnaces lay along the base of the hills in the middle distance

perhaps as early as 14,000 years ago near the end of the Pleistocene Ice Age. These first people would have been attracted by the availability of salt from the natural brine springs and ponds, and they must have hunted the great herds of late Pleistocene mammals that gathered at the salt licks. Thousands of years later in 1787 Thomas Jefferson recorded the first known fossil collected in the Virginia when Saltville resident Arthur Campbell presented him with a "...large jaw tooth of an unknown animal lately found at the Salina" [20]. That "unknown animal" turned out to be a mastodon, an elephantine creature extinct since the Ice Age ended 10,000 years ago.

The salt masses beneath the floor of the Saltville Valley that are the source of the surface brines belong to a geologic unit called the Maccrady Formation. These rocks formed about 350 million years ago in the late Paleozoic Era when the Appalachians were a young rising mountain chain. At that time the continents occupied very different locations from where they are now. Ancient North America lay in the equatorial latitudes and the Appalachian margin was in the trade winds belt. The elevating mountains kept the atmospheric flow of moist air from reaching the Saltville area, thereby forming a dry wind shadow zone along a hot tropical coast. Stretches of the modern Persian Gulf coast have similar environmental conditions that create wide desiccated tidal flats known locally as "sabkhas." In such a setting, the sea water in the lagoons along the shoreline and trapped between the buried sediment grains evaporates, and the dissolved salts precipitate. The two most abundant of these deposits are rock salt and gypsum, both of which proliferate in the Maccrady silts and clays laid down in the long ago Saltville sabkha.

Millions of years after the salt formed, near the close of the Pleistocene Ice Age, natural salt springs, seeps, and ponds dotted the floor of the Saltville Valley. These briny features appear where the fresh ground waters from rainfall and snowmelt circulate down to the depth of the salt beds, dissolve the halite crystals, and flow back to the land surface as saline waters. Attracted to one of the most extensive salt licks in the southeastern quarter of the continent, hordes of mammals migrated into the Saltville Valley. Among the creatures found entombed in the muds of the bogs and small lakes are mammoths, mastodons, musk oxen, giant ground sloths, caribou, moose, deer, and horses. Preying on the huge beasts were short-nosed bears, the largest member of the ursine family that ever existed. Another predator may have been here as well: purported Paleo-Indian artifacts have been found in the late Ice Age sediments with the animal bones.

Following those first people, Native Americans later settled in the valley where they heated the plentiful brines in clay pots to obtain the crystallized salt. White explorers visited the territory in 1748 and in 1753 one of the expedition's members, Charles Campbell, received a patent of land at the Salt Lick in the name of King George II. Campbell's 330 acres contained most of the saline springs and ponds; upon his death the grant passed to his only son William. During the Revolutionary War William Campbell attained the rank of general and commanded the victorious American militia at the Battle of King's Mountain in South Carolina. William's cousin Arthur, presenter of the "large jaw tooth" to Thomas Jefferson, commenced the first commercial development of the salt in 1782. Shortly thereafter other Campbell family members involved themselves in salt manufacture. These original salt works of the late eighteenth century consisted of wells to draw the brine, furnaces in open sheds to heat the saline waters in kettles, and salt houses for storage.

Competition sprang up in 1795 when William King began his own salt procurement on land adjoining the Campbell family. Four years later, King dug a 200-foot-deep shaft intending to mine the buried salt deposits; this is the first known salt mine in the United States. The excavation did not succeed, however, for subsurface brines flooded into the shaft before the rock salt was reached. Unable to overcome the water problems, King reverted to the use of wells and furnaces for salt making. Meanwhile, the original Campbell family operations passed by marriage into the hands of Francis Preston, who retired in 1797 after two terms in Congress to devote full time to salt production. By the turn of the nineteenth century, the Saltville Valley had two commercial entities that provided salt to a region covering parts of five states and beyond. For the next 60 years, the businesses grew and intertwined, becoming known generally as Preston's and King's salt works.

As the decades of the nineteenth century went by, Saltville's salt trade flourished. The War of 1812 brought times of special prosperity when the British blockade created a shortage of salt. Wagons trekked to Saltville from as far away as Richmond and Petersburg to acquire the halite. In the 1820s hundreds of workers toiled in the industry, counting those laboring directly in salt manufacture as well as others involved in supporting businesses. Making and marketing the salt required a host of specialized workers—axe men for cutting timber to fire the furnaces, coopers for fabricating barrels, teamsters to drive the wagons, blacksmiths for the metal shops, and boatmen to raft the product down the Holston River. By 1842, 200,000 bushels (ten million pounds) of salt per year streamed from the complex of facilities in the little valley. Completion of a salt works branch line from the Virginia and Tennessee Railroad in 1856 opened vast unexploited markets, catapulting Saltville into national prominence and setting the stage for its rise to "The Salt Capital of the Confederacy."

In 1857, Harper's Magazine, then a preeminent national publication, sent a writer and artist to Southwest Virginia to prepare an article on the salt works (Figs. 8.2 and 8.3). Saltville had transformed into one of the three biggest salt-making centers in the United States, surpassed only by the enormous establishments at Onondaga and Kanawha. The southwestern Virginia brines had proven to have an exceptionally high concentration of sodium chloride at 98.7 per cent, exceeding those in New York, along the Kanawha, and even at world famous Cheshire, England, the international benchmark for high quality salt production.

In short, Saltville's product was well known in the young America, and Harper's vivid account of its manufacture on the eve of civil war reflected the vitality of the thriving enterprise:

> The salt is procured by sinking wells to the depth of the salt bed, when the water rises within forty-six feet of the surface, and is raised from thence by pumps into large tanks or reservoirs elevated a convenient distance above the surface. The brine thus procured is a saturated solution, and for every hundred gallons yields twenty-two gallons of pure salt.
>
> The process of manufacturing it is perfectly simple. An arched furnace is constructed, probably a hundred and fifty feet in length, with the doors at one end and the chimney at the other. Two rows of heavy iron kettles, shaped like shallow bowls, are built into the top of the furnace – in the largest works from eighty to a hundred in number.

Fig. 8.2 Interior of a Saltville furnace. Note the baskets of salt drying atop the furnace which is located beneath the working floor. Image is from Harper's Magazine, 1857, courtesy of The Library of Virginia

Fig. 8.3 Bags of salt in a "magazine" (storage shed) in Saltville awaiting shipment by rail. Note the off-loaded pile of wood next to the wagon in the left side of the image. The wood was partial payment for the salt. Image from Harper's Magazine, 1857, courtesy of The Library of Virginia

Large wooden pipes convey the brine from the tanks to these kettles, where the water is evaporated by boiling, while the salt crystallizes and is precipitated. During the operation a white saline vapor rises from the boilers, the inhalation of which is said to cure diseases of the lungs and throat.

At regular intervals an attendant goes round, and with a mammoth ladle dips out the salt, chucking it into loosely woven split baskets, which are placed in pairs over the boilers. Here it drains and dries until the dipper has gone his round with the ladle. It is then thrown into the salt sheds, immense magazines that occupy the whole length of the buildings on either side of the furnaces...

The salt thus manufactured is of the purest quality, white and beautiful as the driven snow. Indeed, on seeing the men at work in the magazines with pick and shovel, a novice would swear they were working in a snow-bank; while the pipes and reservoirs, which at every leak become coated over with the snowy concretions, sparkling like hoar-frost and icicles in the sun, serve to confirm the wintry illusion. [3]

This is the technology that produced the Smyth County salt throughout the Civil War. The Confederacy would go to great lengths to defend this nationally important industrial site, and the Union would fight just as hard to destroy it.

Notes

The account of the 1864 Christmas in Saltville is from Stuart [226], a truly riveting eyewitness description of her family's experiences during the raid that destroyed much of the salt works. The remainder of the introductory information comes mainly from Lonn [124], a principal source for much of this chapter and the following one. Holmes [85] also provided material for this chapter, as did Marvel [131]. The section on salt as "An Ancient Mineral Resource" draws most extensively on Kurlansky [115], a comprehensive look at the history of salt from prehistoric to modern times. For the following discussion on "The Growth of the American Salt Industry," I relied on Kurlansky [115] and Lonn [124]. The final part of the chapter, ""Salt...of the purest quality" in Southwest Virginia," begins with geological and paleontological information found in Watson [251], Cooper [33], Ray et al. [190], McDonald and Bartlett [138], McDonald [137], and Sharpe [213]. The Jefferson quote in this part is referenced in Boyd [21]. The concluding discussion of the human history in Saltville is an amalgamation of accounts in Kent [110], Sarvis [199], Marvel [131], Allison [1], the Saltville Historical Society undated, and the Saltville Historical Foundation undated. The lengthy contemporary description of the salt works on the eve of the war from Harper's Magazine is referenced as Anonymous [4] in the Bibliography.

Chapter 9
Two Battles and a Massacre

"They were shooting every wounded negro"

President Abraham Lincoln issued the final version of the Emancipation Proclamation on January 1, 1863, which stated that free black men "...will be received into the armed service of the United States" [119]. That momentous decision inspired General Stephen G. Burbridge, Federal Governor of the District of Kentucky, in 1864 to promulgate General Order No. 24 authorizing the recruitment of freedmen and slaves into black units for service under his overall command (Fig. 9.1). On June 30 white officers assigned to lead the newly instituted 5th United States Colored Cavalry (USCC) obtained permission to begin enlisting volunteers for the regiment. When fully assembled later that fall, nearly all of the soldiers were former slaves, at last able to join the fight for freedom.

In late September, the 600 men of the 5th USCC joined around 5,000 white mounted troops commanded by Burbridge in an invasion of southwestern Virginia intended to decimate the massive salt-making complex in Smyth County. From the start, the black recruits faced deep doubts from their Caucasian leaders and comrades about their willingness and ability to fight. The officers of the regiment had not organized their charges well nor given them anything like proper training. In addition, only second rate weapons and horses had been provided to the 5th. The white troops carried repeating Spencer rifles specifically designed for cavalry, whereas the blacks had single shot infantry Enfield muskets that could not even be loaded from the saddle. Along the route into Virginia, the soldiers of the 5th underwent persistent ridicule, insulting remarks, and stealing of their mounts by their fellow troopers. All of this the black horsemen bore silently with no response; they would wait to prove their mettle on the battlefield.

That opportunity came October 2, 1864, at the First Battle of Saltville. Facing purposefully directed withering fire from Confederate defenders enraged that black men dared fight against them, the 5th USCC pushed slowly up a steep ridge thick with enemy infantry and cannon. A Union officer in one of the white regiments said

© Springer International Publishing Switzerland 2015

R.C. Whisonant, *Arming the Confederacy*, DOI 10.1007/978-3-319-14508-2_9

Fig. 9.1 Union General
Stephen Burbridge. Image
from MOLLUS-MASS
Collection, U. S. Army
Military History Institute.
Burbridge's attack on the salt
works was turned back by
Confederate defenders at the
First Battle of Saltville,
October 2, 1864

prior to the campaign that he did not believe the 5th would really fight, yet during
the battle he had never seen troops contend as hard. The black soldiers carried the
Rebel positions on that deadly slope and hilltop, prompting their commander,
Colonel James Brisbin, to later write "I have seen white troops fight in twenty-seven
battles and I never saw any fight better" [236]. This was the only Federal success in
the engagement, however, and by day's end Burbridge admitted defeat and withdrew back to Kentucky, leaving many of his dead and wounded on the battlefield.

Next day, one of the most appalling atrocities of the Civil War took place at
Saltville. A group of Confederates, led by a violent partisan named Champ Ferguson,
sought out the helpless still-living men of the 5th lying on the ground and merci-
lessly shot them. These brutal executions were not done in the heat of battle, but
rather in the cold light of the following day when the beaten enemy had quit the
field. In truth, an organized band of lawless cutthroats looking for blacks to kill
committed premeditated murder at Saltville in October 1864.

Saltville in the War Years

The Civil War began on April 12, 1861. By that fall the Saltville works had been
acquired by Stuart, Buchanan, and Co. who manufactured salt throughout the conflict
and for a few years thereafter. William Stuart, brother of famed cavalryman Jeb, and
George Palmer were two of the partners in the firm. Palmer had come down in 1858
from the sprawling Onondaga site in New York and taken charge of the operations,
enlarging them significantly by introducing the more efficient salt-making technology
used at Syracuse. (The Syracuse "style" furnaces were exceptionally long structures
in which wood fires at one end generated heat that warmed the salt kettles as it drafted
out through a chimney at the other end.) Shortly after the war started, the business

negotiated a contract with the Confederate government to provide 22,000 bushels (1.1 million pounds) of salt per month for the armed services. Over the next three and one half years, Stuart, Buchanan, and Co. managed to do this and much more.

In the depressed economic times of the decade before the war, the Saltville facilities had dwindled down to only a single furnace and about 70 kettles. At its peak wartime production, the works included 38 furnaces and 2,600 kettles. Union raiders who broke into the valley late in the conflict claimed to have seen as many as 300 buildings making up the physical plant. The huge salt output in the years of fighting, reaching a maximum of four million bushels (200 million pounds) in 1864, commonly exceeded the ability of the Virginia and Tennessee Railroad to transport it. Because the trains could not haul away the white mountains of halite "There would be hundreds of wagons lining the roads for miles waiting their turn for salt. Each wagon would bring a load of wood [to fire the furnaces] to be applied as part payment for the salt and the rest would be paid in Confederate currency" [109].

By the fall of 1862, the Saltville product had become so crucial to the Southern states that Georgia, North Carolina, Tennessee, Alabama, Mississippi, South Carolina, Florida, and Virginia all worked out agreements to purchase salt or erect their own furnaces in Saltville. Soon several state-owned operations went up, most of them along the railroad tracks on the north side of the valley. When it was all over, the prodigious amounts of Saltville salt proved to be of inestimable value to both the national government and the states of the Confederacy. Although shortages happened from time to time, thanks to the Smyth County brines more than any other source, the scarcity was never severe enough to cause serious problems for the military. Confederate Commissary General Lucius B. Northrop noted in January 1865 that "...the supply of salt has always been sufficient and the Virginia works were able to meet demand for the army" [84].

Salt-manufacturing installations of this magnitude could not go on unchallenged by the North, particularly since Union forces controlled most of nearby Kentucky and West Virginia from 1862 onward. In summer of 1863, Colonel John Toland and about 1,000 mounted infantry and cavalry advanced from West Virginia into southwestern Virginia to attack Saltville. That raid came to naught when the Federal leader fell mortally wounded in a sharp firefight at Wytheville and his command withdrew. Another Union foray in September 1863 got within 35 miles of Saltville, only to be turned away after a brief skirmish. The following May, Federal cavalry under General William Averell intended to assault the salt works, but Confederate defenders led by General John Hunt Morgan drove them away in the Battle of the Cove near Wytheville. Now, in the fall of 1864, the most serious Union threat to Saltville so far was assembling over the mountains in Kentucky.

The First Battle of Saltville, October 1864

Saltville in 1864 had emerged as the preeminent salt supply for the Confederate states east of the Mississippi. And yet, as the year lengthened into the fall, no Union expedition had gotten within a few dozen miles of the salt works. That would no longer endure. In September, General Stephen Burbridge, the widely reviled military governor of Kentucky, desperately wanted to eradicate the salines in southwestern Virginia. Although a native Kentuckian, he harshly ruled the border state, meting out reprisals against the citizens for offenses committed by Confederate guerillas, executing or imprisoning many people on false charges of treason or other high crimes, and arresting those he suspected of opposing Lincoln's re-election.

With his reputation worsening rapidly, even among his superior officers, Burbridge needed to show more positive accomplishments. Accordingly, he seized on the notion that an offensive to destroy a major Southern industrial facility would go a long way toward that end. To achieve this aim, he assembled an expeditionary force that totaled about 5,200 men and artillery, including nine regiments of Kentucky mounted troopers and 600 men in the 5th USCC. On September 20, he got underway, bound for Saltville and glory on the battlefield.

Burbridge followed an especially difficult invasion route into southwestern Virginia, moving along the steep slopes of the treacherous, deeply dissected Plateaus country. Traveling over a mountain in a driving rainstorm on the night of September 28, one regiment lost eight riders and their mounts when they slipped from the narrow trail and plunged to their deaths in the canyons below. Others had to be rescued with ropes. Once the invaders entered Virginia, partisans and bushwhackers shadowed and harassed them constantly, effectively cutting off Burbridge from communication with other Union contingents. As a result, when departmental commander William T. Sherman had second thoughts about the expedition and ordered it cancelled, Burbridge did not receive this information and pressed on.

In the meantime on the Confederate side, the newly reorganized Department of Southwest Virginia and East Tennessee assumed responsibility for Saltville's defense. The department's commander, General John Breckinridge, like Burbridge a Kentuckian, had been away campaigning in the Shenandoah Valley and was now hastening back to southwestern Virginia to counter the threat (Fig. 9.2). As darkness

Fig. 9.2 Confederate General John Breckinridge. Photograph courtesy of the Library of Congress. Breckinridge, in overall command of the Department of Southwest Virginia and East Tennessee, arrived in Saltville as the fighting ended on October 2, 1864. His subordinates had won the victory

fell on October 1, Burbridge made camp in Broadford, a scant seven miles north of the salt works. He intended to throw his men, including the 5th USCC, against Saltville's waiting defenders the next day.

Breckinridge had not yet arrived to lead his soldiers in the impending battle, but his chief lieutenant, General John Echols, had been working furiously to pull together every man he could to fight. In Saltville itself, command fell to General Alfred Jackson, derisively called "Mudwall" by his own men, a sobriquet he apparently earned by his ineptness compared to his more famous cousin "Stonewall" Jackson. Nevertheless, this time Mudwall did well in strengthening Saltville's defenses. When the Union columns finally attacked, they found the Confederates stoutly entrenched on the hills guarding the town and salt works.

The evening before the battle General Jackson had an extended conversation with Colonel Robert Preston, leader of a Virginia reserve militia unit in town to help protect the salt operations. As the two talked, the subject of fighting against black soldiers surfaced and this ominous portent of things to come occurred:

> 'Kernel,' said he [Jackson], 'my men tell me the yanks have got a lot of nigger soldiers along. Do you think your reserves will fight niggers?'
>
> 'Fight 'em?' said the old colonel, bristling up; 'by ___, sir, *they'll eat 'em up!* No! not eat 'em up! That's too much! By ___, sir, we'll cut 'em up!' [275]

The First Battle of Saltville commenced on Sunday morning, October 2, 1864. General John Williams came on the field around 9:30 with 1,700 troops just after rifle fire began to crackle (Fig. 9.3). The Mexican War hero, now the senior Confederate officer on the scene, directed Saltville's 2,800 defenders throughout the engagement. Williams arrayed his men along the northern edge of town in a great arc, curving east to west from tall and dominating Chestnut Ridge on the right to

Fig. 9.3 Confederate General John Williams. Photograph courtesy of the Library of Congress. Williams, nicknamed "Cerro Gordo" for his outstanding performance at that battle in the Mexican War (1846–1848), led his men to triumph in the First Battle of Saltville

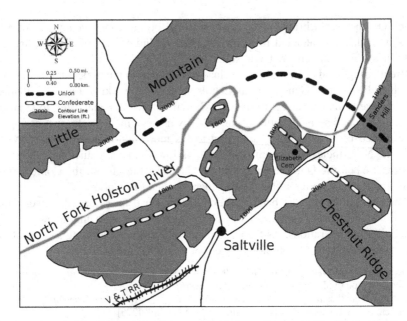

Fig. 9.4 Map of the positions of the Union and Confederate lines at the First Battle of Saltville. The positions are generalized and represent conditions near the start of the battle

Church Hill, a lower rise with a cemetery, in the center (Fig. 9.4). Continuing west, the gray lines ended on the left atop towering bluffs overlooking a ford across the North Fork of the Holston River.

Burbridge opened the action by sending Colonel Robert Ratliff's brigade up and over Sanders Hill to assail the Chestnut Ridge positions. This brigade included the 5th USCC. The Confederate soldiers confronting them were dismounted cavalry belonging to General Felix Robertson's Virginia brigade and Colonel George Dibrell's brigade of Tennesseans. These men had taken cover behind hastily improvised earth, rock, log, and fence rail barricades on the steep hillside. Intense close-quarters combat flared on this flank all day long as the soldiers in blue slowly gained ground, resolutely pushing their opponents up the hill. The 5th USCC fought primarily against the Tennessee unit, many of whom became deeply incensed at the presence of the black warriors. Their musketry focused on these men, who "fell in heaps before the rifles of the enraged Tennesseans" [158].

As the fighting stormed along the slopes of Chestnut Ridge, Burbridge hurled five Kentucky regiments against the Rebel center where Confederate Kentuckians had concentrated on the crest of Church Hill. One of the Federal columns charged across the broad river bottom, taking rifle and artillery fire as they approached the enemy defenses. The Southerners could not hold against this onslaught and fell back, struggling hand-to-hand with their opponents amid the tombstones in the little graveyard on the hill. A Confederate officer galloped off, found some reserves, and ordered them into the combat raging in the cemetery. The re-enforcements rallied the faltering defenders, forcing the Union men to withdraw and saving the Southern center.

While this was going on, Burbridge ordered four additional regiments of Kentucky horsemen to strike the Confederate left, which was anchored by Kentuckians fighting for the South perched on cliff tops rising almost 200 feet above the river. From this commanding high terrain, the men in gray rained down rifle fire on their floundering opponents trying repeatedly though hopelessly to carry the impregnable positions. An almost mirthful attitude developed among the Southerners, a few calling out after a volley "Come right up and draw your salt" [159]. One soldier, after firing at an enemy trooper, shouted "How's that? Am I shooting too high or too low?" [159]. At length the Union brigade leader went down fatally wounded, sending his units into a disorderly retreat and ending combat in the sector.

It was now late in the day, and still the contest continued on bloody Chestnut Ridge. The Union attacks had driven the Confederates into trenches near the hill top, but both sides had little ammunition left by then, reducing many to using their pistols. The sun was setting behind the distant peaks when the last defenders yielded their entrenchments high on the ridge and pulled back toward town. As Burbridge's victorious fighters topped the hill, they could see Saltville and the salt works in the valley below bathed in the fading golden autumn glow. They could go no farther, however. Exhausted and out of ammunition, the Northerners had to turn back in the gathering dusk.

The October 2 fighting in Saltville ceased by 5 p.m. Union thrusts against all of the defensive lines had stalled or failed. Furthermore, three fresh Confederate brigades had come onto the field, intending to resume the battle in the morning. Burbridge meanwhile at last received the order recalling him from the attack, whereupon he turned command over to his senior colonel and headed back to Kentucky. Just as the fighting ended, General Breckinridge arrived and took control of the men in gray. That night, with the other Rebel soldiers, Captain Edward Guerrant watched as "The Yankees built 18 big fires on the side of the mountain to our left, which excited suspicion of their intention to stay" [71]. This turned out to be a deception that effectively shielded the Union forces as they slipped away in the darkness. Thanks to their superb defensive positions, the Confederates lost fewer than a 100 killed and wounded; Burbridge reported a total of 350 casualties.

Next morning, with the battlefield shrouded in dense fog, Guerrant later remembered "We were a little surprised this morning at not hearing the ring of the enemy's rifles making an attack at daylight" [71]. The Union had indeed abandoned the ground, leaving much of the equipment and many of their wounded on the battlefield. Breckinridge sent a brigade after the retreating foe; a few miles north of Saltville they found several wounded men left behind in a cabin, including a few black soldiers. The Confederates maintained the pursuit, traveling back over the Federal invasion pathway while Breckinridge dispatched three more brigades by other routes to cut off the fleeing enemy. Before long the bone-tired pursuers realized they could not entrap the Northerners and gave up the chase. Nearly two weeks afterwards, the Union soldiers reached home in central Kentucky, and Burbridge's campaign to demolish the salt works ended.

The October 1864 engagement at Saltville resulted in a clear Confederate victory, one that saved the salt industry at a time when the South was growing ever more desperate for the resources to keep up the struggle. Although raiders would return in less than three months and do substantial damage to the salt-making facilities, at least for the time being the work went on unimpeded. The Confederate success also meant that the critical lead operations in adjacent Wythe County and the Virginia and Tennessee Railroad running through the region would be secure for a while longer.

The Saltville Massacre

The Saltville Massacre is a tragic and controversial aspect of the October 1864 battle. The first indication of such a horror came the night following the fighting when Confederate Private George Mosgrove, writing in later years, reported that General Robertson told him and Captain Guerrant that "...he thought his men had killed nearly all the negroes" [160]. But the real slaughter of defenseless wounded and captured men began in earnest the next morning. Guerrant described this battlefield scene:

> Scouts were sent, & went all over the field, and the continued ring of the rifle, sung the death knell of many a poor negro who was unfortunate enough not to be killed yesterday. Our men took no negro prisoners. Great numbers of them were killed yesterday & today. [72]

At the same time, Mosgrove noted the discharge of weapons that "...swelled to the volume of a skirmish line" [161]. He rode to where Robertson's and Dibrell's troops still sat posted on Chestnut Ridge; here he found "...the desultory firing was at once explained—the Tennesseans were killing negroes" [161]. Mosgrove heard more shooting and moved forward where he

> ...came upon a squad of Tennesseans, mad and excited to the highest degree. They were shooting every wounded negro they could find. Hearing firing on other parts of the field, I knew that the same awful work was going on all about me. It was horrible, most horrible. [161]

Other witnesses described similar occurrences and claimed to have seen notorious Confederate guerilla Champ Ferguson in particular shooting both black and white soldiers (Fig. 9.5).

More of these shocking acts happened on the Sanders farm, scene of Colonel Ratliff's initial attack the day before. Union Surgeon William Gardner had established a field hospital in one of the outbuildings on the property. Gardner recounted that "...there came to our field hospital several armed men, as I believe soldiers in the Confederate service, and took 5 men, privates, wounded (negroes) and shot them" [235]. In the same vicinity, Mosgrove encountered seven or eight wounded black soldiers in a cabin, lined up with their backs to the walls. He goes on to tell that he saw

> ...a boy, not more than sixteen years old, with a pistol in each hand. I stepped back, telling him to hold on until I could get out of the way. In less time than I can write it, the boy had shot every negro in the room. [162]

Fig. 9.5 Confederate partisan Champ Ferguson. Photograph from Wikimedia Commons. Ferguson was the ringleader of the perpetrators of the Saltville Massacre. He was convicted and hanged for his war crimes in October 1865

Later that morning Mosgrove observed Breckinridge and another general when they came to the scene of the shootings. "General Breckinridge, with blazing eyes and thunderous tones, ordered that the massacre should be stopped" [163]. Nevertheless, as soon as these officers departed, the murders resumed. A few days afterwards, more atrocities occurred in a temporary hospital at nearby Emory and Henry College where Confederate and Union wounded were being treated. On October 7, Champ Ferguson and another Rebel partisan entered the building and shot two black soldiers in their beds. Ferguson returned the next day and killed a white Union officer, Lieutenant Elza C. Smith, where he lay. Breckinridge later reported these crimes to General Robert E. Lee, accusing Robertson of having directed the killings. The war ended the next spring, however, without anyone being held accountable for the murders at Saltville.

The total count of black troopers executed at Saltville is not known with certainty. After the battle, the chief surgeon of Burbridge's division filed a casualty report stating that for the 5th USCC, 22 men were killed and 96 wounded or captured. Estimates of the slaughter range from as few as five to 100 or more. The most extensive study of the massacre gives a "conservative" number of 46 black soldiers murdered, noting that Saltville stands as "…possibly the worst battlefield atrocity of the Civil War" [132].

The Second Battle of Saltville, December 1864

By December 1864 the tottering Confederacy's ability to protect its territory had deteriorated markedly. Remote places such as Southwest Virginia had been nearly stripped of defenders, the manpower more acutely needed for the main fronts. Many of the men that fought to save the salt works in October had departed, leaving General Breckinridge with fewer than 2,000 regulars to safeguard the area. Union General George Stoneman, now the acting commander of the Department of the Ohio, decided to strike again at Saltville.

On December 10, Stoneman left Knoxville with about 6,000 mounted troopers and four artillery pieces, headed for southwestern Virginia. He aimed to wreck the salt installations, the Wythe County lead mines and smelters, and as much of the Virginia and Tennessee Railroad as possible. Brushing aside the few Southerners standing in the way, the Union expedition moved swiftly up the Great Valley, leaving in its wake ruined railroad stations, rolling stock, and tracks between Bristol and Wytheville. On December 17, a detachment of Stoneman's command overran the lead mines, then wheeled back south toward Saltville. Next day, the Federal horsemen met and swept aside Breckinridge's meager units at Marion. The way to Saltville lay open.

The Northern troopers arrived at the southern approaches to Saltville on December 20 and the second battle in less than 3 months broke out about 2 p.m. (Fig. 9.6). Confederate cannon fire from Fort Breckinridge atop a tall hill, supported by batteries on the ridges bounding the north side of the valley, held up one of the

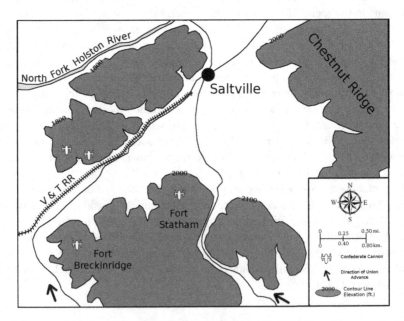

Fig. 9.6 Map of the Second Battle of Saltville, December 20, 1864. Note the approach of the Union from the south through the gaps in the hills. The Confederate cannon were located in earthen fortifications

Union columns until night fell. Moving stealthily in the darkness, Federal Colonel Brazillah Stacy led his 1st Tennessee cavalry past Fort Breckinridge, then had his charges dismount and walk their horses up the steep slope behind the earthworks. Cresting the hill top, they remounted and drove off the surprised Confederates. Stacy sent some of his riders back downhill into the Saltville Valley to begin demolition of the salt works. The colonel advanced with the remainder to a second ridge top fortification, Fort Statham, only to discover it already vacated by the panicked defenders.

Seeing that his two principal defensive positions were lost, the Confederate commander in Saltville evacuated the town, leaving the salt operations completely in the hands of their captors. In the early morning hours of December 21, the rest of Stoneman's raiders rushed into the valley and set about demolishing the mineral-producing facilities

> ...in an orgy of destruction. Sledge hammers rang against salt kettles and the masonry kilns; artillery shells and railroad iron rattled down the wooden well casings; soldiers broadcast sacks of salt like Romans at Carthage; everywhere sheds, stables, and offices crumbled in flames. [131]

Despite their earlier defeat at Marion, Breckinridge's troops had cautiously proceeded toward the salt works, hoping to somehow rescue the vital facilities. Late in the day on December 20 they neared Saltville where reconnaissance found the Union forces blocking the way into the valley too strong to confront. In the evening gloom Captain Guerrant, a veteran of the October clash, saw the distant light of burning Saltville. The distraught Confederate officer described his feelings:

> BURNING SALTVILLE! It was done! Finished! 'love's labors lost,' & our tears in vain... Goodbye dear Saltville! We have marched for you, watched for you, fought for you, bled for you, *died* for you for these many years, but you are now as dead as Doesticks friend! Farewell! [73]

Their work done, Stoneman's men left Saltville and withdrew from southwestern Virginia. The "orgy of destruction" notwithstanding, salt making had not been shut down for good. The immense industrial center was simply too much for a few thousand soldiers to permanently disable in a matter of hours. A report to Breckinridge shortly after the Saltville raid stated that less than two-thirds of the sheds and one-third of the kettles had been destroyed; some of the buildings and furnaces were left untouched. Several weeks later workmen had the operations running again and furnishing salt for the Southern states and Confederate government.

In April 1865, as the Confederacy collapsed, Stoneman returned to Southwest Virginia to devastate the lead mines and railroad once again, but chose not to attack Saltville this time. By then, no amount of salt or lead or any other mineral resource could save the exhausted South. Grant accepted Lee's surrender at Appomattox on April 9, even as workmen at Saltville faithfully turned out another day's run of the snowy crystals.

Notes

The citation from Lincoln [120] beginning this chapter I found online at the National Archives & Records Administration website. The rest of this introductory section is an overview of the first battle of Saltville and the massacre that followed summarized from Davis [42], Marvel [129], Marvel [131], Mays [134], and Mays [136], five references central to understanding these topics. Colonel Brisbin's comments regarding the valor of the black troopers during the fighting are from the *Official Records* [236]. For the discussion of "Saltville in the War Years," I used Lonn [124], Rachal [185], Kent [110], Sarvis [199], Holmes [85], and the Saltville Historical Foundation (undated). The detailed accounts of "The First Battle of Saltville, October 1864" and "The Saltville Massacre" are, once again, from Davis [42], Marvel [129], Marvel [131], Mays [134], and Mays [136]. Eyewitness descriptions of the combat and subsequent murders of the black soldiers from Surgeon Gardner as recorded in the *Official Records* [235], Wise [276], Mosgrove [164], and Guerrant [74] also make up a substantial part of these two sections of the chapter. The concluding narrative, "The Second Battle of Saltville, December 1864," is largely derived from Marvel [131], but Lonn [124] and Guerrant [74] also provided information.

Chapter 10
Iron, Civilizations, and War

"Cold Iron – is master of them all"

In 1841 a young engineer from western Virginia quit his job building turnpikes for the state and accepted appointment as chief agent for a struggling iron works in Richmond. Within a year Joseph Reid Anderson secured contracts for the Tredegar Iron Company to make cannon, shot, shell, and chain cable for the United States Navy (Figs. 10.1 and 10.2). This initiated a 22-year period when the firm never lacked orders for big guns as well as a multitude of smaller armaments. In the Civil War, Anderson's firm went on to become the bulwark of the Confederate ordnance program and its single most important industrial operation.

The South had to have Tredegar and the Richmond area's numerous other forges and foundries to arm its military forces; that was one of the main reasons why the Confederacy early on shifted the capital from Montgomery, Alabama, to Richmond. Virginia in fact constituted "…the steel helmet of the Confederacy…" [15] and its advanced weapons industries simply could not be lost. Therefore, the relocated Confederate seat of government, though not far from Washington and sure to be attacked fiercely and repeatedly by Union armies, gained much more security by moving to a state and city that would be the most staunchly defended bastions to the very end.

At 4:30 a.m., April 12, 1861, a Rebel battery lobbed the opening shot of the war into the night sky above Fort Sumter and Virginia's iron industry was there. The Tredegar Iron Works cast the 10-in. mortar that fired the signal round to open the cannonade. Of the 48 artillery pieces that bombarded Fort Sumter, that same company made at least 20. By the end of hostilities in 1865, Tredegar—highly dependent on raw iron from western Virginia—had turned out almost half of the cannon and about 90 per cent of the artillery ammunition used by the Confederate armed services. Despite overwhelming wartime problems with shortages of labor and raw

© Springer International Publishing Switzerland 2015

R.C. Whisonant, *Arming the Confederacy*, DOI 10.1007/978-3-319-14508-2_10

Fig. 10.1 Confederate
General and Tredegar Iron
Works owner Joseph
Anderson. Photograph
courtesy of the Library of
Congress. Taking over
Tredegar in the 1840s, by
1860 Anderson had built the
Richmond firm into the
largest manufacturer of iron
products in the South

Fig. 10.2 Part of the Tredegar Iron Works along the banks of the James River in 1865. The quantity and quality of Tredegar's ordnance and munitions output made it indispensable to the Confederate military effort. Photograph courtesy of the Library of Congress

materials, an underdeveloped transportation system, and outright meddling from government officials, Joseph Anderson's factories and mills out-produced every heavy ordnance maker in the highly industrialized North save one, the giant R.P. Parrott Company in New York. Tredegar by and large forged the Southern iron sword of war, and the bright, energetic Anderson was the mind and heart of Tredegar.

The World's Most Precious Metal

Iron is the most widely used metal in human history. It was essential for conflict in the 1860s, as it had been for the previous 3,000 years. Kipling caught the fundamental value of iron in the heyday of the British Empire in these lines from *Cold Iron*:

> Gold is for the mistress – silver for the maid –
> Copper for the craftsman cunning at his trade.
> Good! said the Baron, sitting in his hall.
> But iron – Cold Iron – is master of them all [196].

Iron is the fourth most abundant element in the earth's crust, making up five per cent of the planet's outer layer. Besides being exceedingly common, iron is also very difficult to smelt because it typically occurs tightly bound to other elements. The unusually high temperatures needed to extract the pure metal caused any widespread use to lag far behind that of copper, bronze, lead, and a number of other more easily refined metals and alloys. Beads dated around 3500 BCE and composed of iron from meteorites, where it is found in a naturally metallic (uncombined) state, are known from Egypt. The Pharaohs considered metallic iron to be more valuable than gold, at least in part owing to its occurrence in objects fallen from the skies. Such an origin signified being sent from the gods and therefore marked the metal as "holy."

In China the knowledge of iron may go back to 4000 BCE. Poets in India said that their ancestors used iron, and objects from smelted iron occur there as early as 1800 BCE. By 1000 BCE the technology of advanced iron processing traveled to China from the West, and 500 years later cast iron objects, including large kettles, appeared there. In a few centuries, Chinese iron masters had invented early versions of blast furnaces that used bellows, some of them driven by water wheels. Shortly after the onset of the Common Era, the Han dynasty built clusters of blast furnaces using a bellows system of forced air to raise the temperatures significantly. Chinese iron smiths had long since achieved improvements in steel making and fashioned the resulting material into exceptional tools and weapons. Realizing the importance of keeping such an innovative ferrous industry secret, officials decreed iron (along with salt) to be a government monopoly. Western steel makers did not rediscover some of the Chinese techniques of introducing carbon into molten iron until the eighteenth and nineteenth centuries.

Among the earliest smelted iron objects in the archaeological record is a dagger blade forged about 2500 BCE in the Anatolian region of Turkey. Some biblical scholars believe that the story of Cain and Abel as recounted in Genesis is a morality tale about the malevolence of those who made iron and used it in warfare versus the goodness of a simple pastoral society with no such weapons. Cain, or kayin in Hebrew, means "smith," as in one who works with metals. Tubal Cain referred to a territory and its inhabitants along the shores of the Black Sea in northern Turkey and thus provided the name of the legendary first iron weapons maker. The ancient Israelites, being relatively peaceful nomadic shepherds, perhaps thought of ferrous objects for killing as inherently evil. The Genesis verses, then, are the story of how the murderous iron user Cain slew the virtuous herder Abel with a forged edged instrument like a dagger or knife. Regardless of possible biblical connections, northern Turkey, the land of Tubal Cain, was indeed a primary center of metals production, especially bronze. That expertise in processing copper and tin might have led directly to the successful purification of iron from its ore minerals in the same locale.

Smelted iron artifacts made an early appearance elsewhere in the ancient world, as for example, a dagger buried with Egyptian King Tutankhamen in 1400 BCE and a battle axe from Syria dated 1450–1350 BCE. The Hittites, an Anatolian people who rapidly rose to mastery of the region in the late Bronze Age from 1500 to 1200 BCE, had extensive knowledge of extracting and forging iron from its commonly occurring ores. They ranked among the earliest to equip their military forces with ferrous weapons, most prominently daggers and swords that greatly helped them to overcome opposing armies still using bronze arms in battle.

Iron had several crucial advantages over bronze—it was stronger, more flexible, and could hold a sharpened edge longer. In addition, the dulled iron weapon edge could be honed back to sharpness, unlike a bronze implement that had to be entirely recast. Perhaps most important, iron deposits are bountiful, and this abundance drove the early iron makers to at last succeed in developing efficient metallurgical techniques to extract the metal. Once this happened, ferrous objects of all kinds became far cheaper to produce than bronze, which required less plentiful copper and relatively rare tin to make.

Learning the art of refining iron ore into strong and useful objects did not come easily, however. This difficulty for centuries gave iron far more value than even precious metals; at times in Assyria, its price exceeded that of gold by a factor of 40. And once smiths could readily extract metallic iron, challenges still remained. Its brittleness in the initial crude metal weapons caused them to shatter in combat, a defect that took time to overcome.

Iron is found on Earth almost always as a chemical compound with other elements, most often oxygen (as in common rust, iron oxide), and melting out the metal demands high temperatures of 2500 °F or more. This much heat far exceeded that needed to smelt copper and tin, but ongoing advances in Bronze Age furnaces eventually enabled them to reach such temperatures. The iron produced from those more progressive furnaces came out in separate dense masses, albeit solid ones rather than liquid. Likely discovered by accident, metal workers found that hammering this malleable clump of crude iron forced out the slag, leaving behind a much more refined material called wrought iron. Repeated heating and hammering plus experiments involving the addition of different elements like nickel steadily increased the strength of the iron.

The most critical additive of all—carbon—transforms iron into steel at low percentages. Exposing iron to air draws in carbon and once better furnaces gave forth molten iron, ancient smiths discovered that careful stirring of the liquid and mixing in air worked to create steel. Around 1100 BCE iron masters found that quenching hot iron implements in water made the metal harder. Improving these processes, in particular learning the correct amounts of carbon to mix in to generate steel, solved the early brittleness problem. (But the amount of carbon must be tightly controlled and other steps involving quenching and tempering are necessary. None of this was well understood until the mid-nineteenth century when Henry Bessemer conceived his method of making high-grade steel inexpensively and efficiently.) Within a century, ferrous weapons, tools, ornaments, and a host of other objects became as common as bronze ones, in time nearly completely replacing them.

As iron forgers upgraded their product, the new techniques of working the metal spread swiftly around Asia Minor and the eastern Mediterranean. Diminishing supplies of good quality copper and tin ores added to the quickening pace of transition from the Bronze Age to the Iron Age. Iron manufacture grew into substantial industries in Cyprus and Greece by 1000 BCE, disseminating from there along the Mediterranean coast and into northern Europe. The expansion of this technology into central and northern Europe linked directly to the success of Gallic warriors whose smiths provided them with very good iron weapons, most notably swords.

Around 390 BCE the Gauls, armed with these better tools of combat, routed a Roman army and pushed on to pillage Rome itself. Their vanquished Latin opponents soon changed over from Bronze Age weapons and tactics that had brought defeat to the newer ways of waging war, thereby beginning the ascent to dominion that lasted for centuries. A Roman incursion into England in 54 BCE under Julius Caesar found iron works already established south of present-day London. This part of the British Isles remained a center of production over the many years of Roman rule.

Extensive use of iron took place throughout the Empire; skilled forgers made a host of items, including nails, hinges, bolts, keys, chains, weapons, and armor. When legionnaires withdrew from one outpost in Scotland, they left behind seven tons of iron nails, a sure indication that the metal was cheap and abundant. The ferrous implements of fighting devised in ancient and classic times—swords, spears, arrowheads, battle axes, shields, and armor—remained essentially unchanged until the rise of gunpowder firearms 2,000 years later.

The Roman Empire collapsed in the late fifth century, but iron making went on in Europe and the product continually improved, particularly in Spain. Construction of the earliest blast furnaces on the continent took place in Germany, Switzerland, and Sweden in the twelfth and thirteenth centuries, making iron implements much less expensive and widely available. A kind of energy crisis took place in England in late medieval times due to the vanishing wood supply. By then, axe men had chopped down many of the old growth timber stands, including the original Sherwood Forest, for charcoal to feed the ravenous furnaces and forges. When English settlers arrived on the shores of America in the early seventeenth century, iron manufacture had already lessened in the homeland because of the shortage of wood for charcoal. Regardless, the growing British Empire demanded still more iron, setting the stage for the formation of a colonial industry that would greatly help supply the Mother country, then ultimately become the principal source for the entire globe.

Iron in America Before the Civil War

The first English colonists ventured to North America hoping to find immense amounts of gold and silver as the Spanish had farther south; however, there would be no such discoveries immediately. Even so, a strong mineral industry did spring up, one that quickly spread throughout the new settlements—the production of iron. An earlier expedition in 1585 led by Sir Walter Raleigh had found iron in what would become North Carolina, but nothing came of that discovery. Iron ore mining originated in America at Jamestown in 1607 and the initial smelting furnace appeared near Richmond around 1620. By the end of the seventeenth century iron manufacture had started up in most of the other colonies.

These newly begun operations used a soft, spongy kind of iron-rich sediment called bog ore. Despite the fact that this was a relatively low grade material, it abounded in the coastal wetlands along the eastern seaboard and could be converted into reasonably good metal. Moreover, vast forests provided plentiful wood for charcoal and the loose seashell accumulations near the shores made excellent lime flux to draw slag from the iron. Small and inefficient though they were, the beginning colonial furnaces could bring forth solid lumps of iron called blooms from finely crushed ore. As had been done for thousands of years, smiths then worked the blooms by heating and hammering until wrought iron resulted. Using such time-tested methods, a typical American bloomery could turn out 12 to 20 pounds of iron a week. Most of the metal stayed in the local vicinity, going into items like pots, skillets, kettles, wagon wheels and tools of all kinds essential for life on the frontier.

The commencement of large scale commercial iron making took place along the banks of the Saugus River in Massachusetts Bay Colony in 1646. A small works there named Hammersmith evolved into a complex of different facilities, including a blast furnace, forge, foundry, and trip hammer. The Saugus itself offered a source of power to a number of water wheels and a means to transport the iron materials to market. A rolling and slitting mill, the earliest in the Americas, made iron plate for cutting, or slitting, into rods to form nails and wire. After only 22 years in business, rising labor costs and ownership issues forced Hammersmith to close. Nevertheless, it was a harbinger of the massive American iron and steel industry to come.

Additional seventeenth century iron works sprouted up in the colonies, notably at New Haven, Connecticut, in 1658 and Pawtucket, Rhode Island, in 1675. New Jersey's industry started about 1665 and by the end of the next century, one county alone had ten mines sending ore to two furnaces, three rolling and slitting mills, and 40 forges. The New Jersey furnace at Batsto obtained bog ore from the nearby Little Egg Harbor swamps to make cannon and shells for the Revolutionary army. Maryland bog iron gave rise to operations that began shortly after 1681. This state was a leader in iron output in the late 1700s prior to the Revolution. After American colonists won independence, Baltimore transitioned into a significant iron-making city, thanks to copious amounts of ore, limestone, timber, and water power in the area. New York also experienced a surge in iron manufacture after prospectors located major ore deposits in the northern part.

Of all the 13 colonies, Pennsylvania created the most extensive and long-lived iron industry, one that became intimately associated with coal extraction and an invaluable asset to the Union war effort in the 1860s. An early iron works appeared around 1720 and in 1742 a furnace went into blast at Cornwall using ore from mines close by. Cornwall achieved great prominence based on the exceptional 70 per cent iron content of its ore. At its peak in the nineteenth century, this sprawling "iron plantation" encompassed numerous industrial, residential, and agricultural activities. The last Cornwall furnace closed in 1883; however, the mines stayed active until 1973, making them the oldest continuously operating mines in the Western Hemisphere. Another famous Pennsylvania iron site was the old Valley Forge location where Washington and his Continental Army passed the dark and dreary winter days of 1777–1778. The general and his men crossed the Delaware River to attack Hessian mercenaries on Christmas night of 1776 using special forty-foot-long boats designed for transporting iron ore.

Ultimately, it was western Pennsylvania iron and coal that lifted America's iron and steel industry to its grandest heights. Coal mining got underway in the Pittsburgh area in 1784 and an iron furnace began operating six years later. Located where three great rivers merge and with thick bituminous coal layers all around, Pittsburgh transformed into the premier iron and steel locus in the United States for the next century and a half. On the other side of the state, Philadelphia put together its own network of furnaces and forges that made it a major exporter of ferrous products by the late 1700s.

Although each of the Southern colonies generated some iron at one time or another, the principal centers stayed north of the Potomac, first in Massachusetts, then New Jersey, New York, Maryland, and finally Pennsylvania. That lead was never relinquished and gave the Union a huge advantage against the Confederacy when industrial strength became decisive in the long, grinding sectional fight to come. Yet an antebellum iron industry did exist in the Southern states to build upon, nearly all of it in Virginia. Iron mining and smelting in America started up here, but then the state's output declined for many years. In the decades before 1860, Virginia iron making revived and expanded once again, this time centered around a core of companies in Richmond. When war came, the capital city's heavy industry would be indispensable to a Confederacy fighting for survival.

Development of Virginia's Iron Industry

Enterprising settlers set about digging coastal bog ore in the first year at Jamestown, shortly thereafter exporting 17 tons to England that yielded "superior metal." A furnace went up near present-day Richmond around 1620, only to be wrecked by an Indian attack in 1622 that spelled the effective end of iron making in the colony for nearly 100 years. In 1716 Governor Alexander Spotswood re-established iron manufacture when he erected a blast furnace near Fredericksburg able to put out molten pig iron, the earliest such operation in the country. Over the next decades iron operations sprang up here and there in eastern Virginia, but a significant industry failed to take shape. By then it was quite evident that growing crops, above all tobacco, in the fertile bottomlands offered a much faster way to wealth than the more time-consuming and expensive mining, smelting, and forging of iron.

The virtual abandonment of the development of iron works in eastern Virginia left the way open for the emergence of a much longer-lived industry located in the Great Valley just beyond the Blue Ridge. Hardy pioneers of German heritage mostly from Pennsylvania trekked south along this natural corridor, arriving in the Shenandoah Valley as early as 1728; Scots-Irish folk followed not long after. Early in the 1740s a Pennsylvania Dutchman had an iron furnace going in the northern end of the Valley and more such endeavors spread farther south during the next 20 years. Thomas Jefferson recorded in his *Notes on the State of Virginia* in 1781 that works in the Shenandoah country provided over 1,000 tons of pig iron a year. By then, conflict with England had resurrected the iron industry throughout Virginia, and fires blazed again in furnaces and forges east and west. Nonetheless, the industry had become well rooted behind the Blue Ridge and would persist far longer than the operations toward the coast.

In April 1775 "the shot heard round the world" at Concord ushered in open warfare between England and her upstart American settlers. Three months later, an assembly of Virginians convened and determined to step up iron production in the colony. The decision to build a small arms factory at Fredericksburg resulted, and a year and a half later the little armory began turning out gun barrels, ramrods, and bayonets. In 1779 a foundry launched operations near Richmond that generated, in addition to various tools and implements, cannon balls, grape, and canister shot. The war years brought forth more iron works, most of them concentrated in the Piedmont and Great Valley. Within a year of the British defeat at Yorktown, reduced demand for iron caused the shutdown of many Piedmont enterprises, but things were different in western Virginia. Survival in the mountain wilderness depended on a steady supply of iron tools, household items, and weapons, thus the number of mines, smelters, and forges steadily increased.

In the early nineteenth century the state of Virginia constructed an arms factory at Richmond that, according to an 1810 report, came to be one of the three largest cannon foundries in America. The artillery plant had two furnaces, one each for casting bronze and iron cannon. Over the mountains, iron works in the Great Valley churned out metal famous for its high quality from Virginia to New England. Barges floated much of this product down the James River to foundries in the Richmond area, establishing a firm foundation for the heavy industry that even then was arising in the capital. Improvements to the Federal armory at Harper's Ferry and the ongoing demand for iron from the state-owned installations at Richmond drove much of the expansion of the western operations. By then iron making had spilled out of the Shenandoah Valley farther and farther into the remote southwestern reaches of Virginia.

In the 1830s Richmond showed promise of becoming a nationally renowned metal manufacturing center. The city had the James River to provide water power to the big factory machines and to carry away the products to markets. Several firms fabricating iron or brass devices were there, including the Tredegar Iron Company created by an act of the Virginia legislature in 1837. Another busy iron endeavor in the capital built a locomotive for the Richmond, Fredericksburg, and Potomac Railroad. In addition to the metals works, flour mills, tobacco factories, paper plants, and cotton and woolen processing facilities all started business in Richmond in the antebellum years.

Virginia's industrial base grew rapidly, propelled in large part by the iron and coal industries at Richmond. Some predicted that the capital would become "the Manchester or Birmingham of America," and in fact it was on course to do just that. In 1840 the Richmond and Petersburg area hosted 42 iron furnaces, 52 bloomeries, and a number of forges and rolling mills that provided over 6,000 tons of bar iron. Coal emerged as a primary fuel and the prolific Midlothian mines just west of Richmond extracted enough to rank Virginia second in the country behind Pennsylvania. Meanwhile the flourishing iron works beyond the Blue Ridge continued to be the principal source of raw metal for the forges and foundries in the east.

In the decade before the Civil War, an enormous change came over Virginia and the rest of the nation—railroads expanded tremendously. Explosive growth of the rail system in Pennsylvania enabled the state's anthracite-produced iron to flood the market with cheap metal of good quality. Many of Virginia's charcoal furnaces could not compete and fell out of blast. Iron manufactured using charcoal cost twice as much to make as that from anthracite, yet many iron masters regarded the charcoal product as the better choice. Owing to this ongoing market for top grade charcoal iron, those Virginia iron works putting out the best metal prospered against the northern goliath, and most lay west of the Blue Ridge.

The 1850s saw the Tredegar Works in Richmond rise to the top of the big establishments turning out finished iron products. Joseph Anderson, at first the sales agent for the firm, had purchased it outright in 1848. The new owner, educated at West Point where he roomed together with Thomas J. Jackson for a year, proved to be an excellent businessman. Prior to 1850, Tredegar sold most of its output in the North, but afterwards Southern customers dominated as Pennsylvania iron gained tight control of the markets north of the Potomac.

Sales in the South were driven in part by the escalating iron demand from the region's expanding railroads. The far-sighted Anderson had seen this trend and soon his company was building locomotives and associated rail hardware such as spikes, chairs (used to hold the rails together), axles, and bridge iron. By the end of 1855, at least 41 Tredegar locomotives rolled along Southern tracks. Some of the corporations running the railroads insisted on Anderson-made equipment because of its consistently outstanding quality. The management of the Virginia and Tennessee in southwestern Virginia would accept no other besides Tredegar axles for the company's passenger and freight cars.

Manufacturing iron for railroads was indeed good business, but Anderson diversified his operation into many other areas, among them making metal for nuts and bolts, bridges, ships, and a variety of industrial equipment. Soon after taking over Tredegar, he plunged deeply into another promising market—military ordnance. By 1860 over 800 guns had been cast for the government in Washington and another

64 cannon for the state of South Carolina. During those years, the firm also made machinery and other components for two new United States Navy frigates. In 1859 Anderson merged Tredegar with another Richmond business, the Armory Iron Works, dramatically enlarging the size and scope of his operations. Around him, several other noteworthy iron manufactories thrived, the Belle Isle Manufacturing Company and Richmond Iron and Steel Works being prime examples. Railroad products, agricultural equipment, and military ordnance poured from the Virginia heavy industry factories to supply the surging Southern economy.

As war clouds gathered, no city in the South rivaled Richmond's ability to put out huge amounts of iron for a multitude of uses. In 1860 the capital had four rolling mills, 14 foundries and machine shops, a nail factory, six businesses making rails, two circular saw shops, and 50 other iron and metal operations employing 1,600 mechanics and nearly 3,000 additional workers. That same year, Tredegar signed a contract with the state of Virginia to provide equipment to fabricate 5,000 muskets annually. In April 1861 Virginia seceded and took over the United States arsenal at Harper's Ferry before completion of the Tredegar project with the state. Harper's Ferry had a sizeable store of weapons, particularly rifles, thereby decreasing Virginia's need for muskets. Consequently, the Confederate government paid for the machinery already made, then transferred the rest to the old Richmond armory where it eventually formed the core of the Confederate Ordnance Department.

Since the fall of 1860 Anderson's iron works had been filling orders from South Carolina, Georgia, and Mississippi for cannon, shot, shell, gun carriages, limbers, and caissons. Tredegar kept busy casting heavy ordnance for Virginia and Alabama as well. Although Alabama had substantial reserves of iron ore, the state was just starting to develop its own iron industry. Wartime demands spurred the shops and mills at Selma to grow quickly and the state's output exceeded Virginia's before the conflict ended. Despite the rapid expansion, Selma did not begin to manufacture cannon until 1863.

When hostilities broke out, the South required heavy weapons immediately and only Virginia could make those. Moreover, at that time the Virginia state armory in Richmond alone in the Confederacy had the capability of producing muskets. Another critical asset was the Federal navy yard at Norfolk, seized by the state just after Fort Sumter surrendered. This gigantic facility possessed a considerable quantity of military stores and equipment, including a foundry and boiler shop that the Southern armaments industry badly needed.

This invaluable war-making capacity and the stream of raw materials coming from the iron, lead, salt, coal, and niter works marked Virginia as the one state the Confederacy had to have. The advanced heavy industry concentrated in Richmond stamped it as the one city above all that must not fall. The momentous transfer of the Confederate government from Montgomery to Richmond in May 1861 showed clearly that the struggle for Virginia and its capital would be bloody and to the death. Lee, Jackson, and the Army of Northern Virginia would go on to great glory and legendary status, but from the beginning wise Confederate leaders knew that only Virginia's mineral resources and manufacturing based on those resources could truly arm and sustain the South in a prolonged industrial conflagration.

Notes

The opening section on Joseph Anderson and the Tredegar Works is taken from three sources: Bruce [28], Norville [171], and Crews [37]. Kathleen Bruce's book, *Virginia Iron Manufacture in the Slave Era*, is the fundamental reference for this chapter and the one following; it is still the most detailed recounting of the rise of Virginia's iron industry and its importance during the Civil War. The "World's Most Precious Metal" discussion is built on Craig et al. [36], Jensen and Bateman [95], and the Cowen [35] website chapter on iron. The interpretation of the Cain and Abel story as evil iron maker versus virtuous farmer I found in Asimov [10]. The story of "Iron in America before the Civil War" draws largely from St. Clair [219] with some information from Mirsky [153]. The ending section on the "Development of Virginia's Iron Industry" is derived entirely from Bruce [28].

Chapter 11
Virginia's Iron Industry in the Civil War

"The iron was wanted more than anything else but men"

Union General George Stoneman's December 1864 raid cut a wide path of destruction in southwestern Virginia. That attack, intended primarily to disable the salt works, lead mines, and Virginia and Tennessee Railroad, also found and leveled some of the iron-making sites in the Great Valley.

Farther north, Federal commander David Hunter had earlier devastated much of the southern Shenandoah Valley in his June 1864 campaign against Lynchburg. Known as "Black Dave" for his harsh treatment of Southerners and their possessions where his forces passed, the general had once again lived up to his nickname. Driving toward Lynchburg, Hunter happened upon some iron furnaces and forges, several of them primary suppliers to Tredegar, and left them smoking ruins.

No Union campaign ever targeted the western Virginia iron operations as the single operational goal; the sites were too numerous and too scattered to be the focus of an offensive against them. Even so, many of the Confederate furnaces still operating in 1864 resided in Virginia west of the Blue Ridge where they were vulnerable to Federal raiders like Stoneman and Hunter. The collective output of these sites had provided the basis upon which Virginia's pre-war heavy industry was built and, now that war had come, played a key role in keeping the Rebellion alive. The loss of any of them was a loss the Confederacy could ill afford.

Confederate Iron and the Tredegar Works

Societies wage war based on their resources—manpower, financial, manufacturing, and raw materials. In early 1861, Virginia had the largest population, most varied production base, and most developed minerals industry in the slave-holding section. In the deeper South, the elitist plantation grandees had maintained the political

© Springer International Publishing Switzerland 2015
R.C. Whisonant, *Arming the Confederacy*, DOI 10.1007/978-3-319-14508-2_11

mastery necessary to keep their cotton-centered agricultural enterprises firmly in control of the state economies. But in Virginia a more or less informal coalition of planters, businessmen, bankers, industrialists, and politicians had formed, one that envisioned a much broader commercial foundation that could compete more effectively with the North.

Starting in earnest in the 1830s, initiatives by this group and other progressive entities resulted in a spurt of road, canal, and railroad building that had an immediate and enormous impact on the state's economy. A major outcome was a burgeoning transportation web connecting the natural resource-rich west with the industrializing east and its investors looking for different ways to enhance their financial wealth. As the rail lines expanded and farms and factories needed more machinery, the demand for iron skyrocketed. Richmond stood at the epicenter of iron making, thanks to its long history of metal output founded on water and coal fields at hand to power the mills and a ready supply of raw iron from the western valleys. At the same time, Petersburg, Lynchburg, and Fredericksburg also emerged as important manufacturing cities where iron operations contributed materially to the modernizing economy.

Although Virginia claimed the most extensive iron industry in the South, other states in the section at least had the beginnings of their own. Alabama in particular possessed considerable reserves of ore and abundances of wood, water, and limestone. Utilizing these assets, forward-looking entrepreneurs at Selma and elsewhere across the state erected furnaces and forges in the antebellum years. In the course of the hostilities, the Alabama operations managed to overtake the Virginia producers; in the end, the two states generated about 90 per cent of the Confederate iron. Smaller iron works existed in five other Southern states—North and South Carolina, Georgia, Tennessee, and Mississippi. In addition to the Virginia factories, ten rolling mills functioned in the South in 1859, one of the biggest being the Gate City Rolling Mill in Atlanta. By 1861 Gate City was putting out armor plate for the new ironclads along with railroad iron and other items essential to the prosecution of the war.

In spite of these domestic iron providers, the Confederacy never had enough; the amounts needed for fighting the industrial conflict now upon it were just too much for the South's economy to provide. The metal was critical for the armaments factories to be sure, but railroads consumed huge volumes as well. The Confederate tracks and rolling stock began the war much inferior to the Union's, and soon a persistent lack of iron made simply keeping the lines running the most pressing problem. Maintaining and extending the railroads required more than twice the amount of rolled steel yearly that the combined productive capacity of all the Southern furnaces could muster.

To make the effect of scarce iron even worse, unanticipated uses such as plate for the ironclad gunboats profoundly increased demand as the struggle went on. In late 1862 Ordnance Chief Josiah Gorgas reported gravely that "Our contractors are many of them at a dead stop for lack of iron, and it will be impossible to supply projectiles for the new guns now in the *field*, unless it can be had" [241]. Concerned about the seriousness of the shortages in 1863, John Jones, an exasperated government clerk in Richmond, noted in his diary: "The iron was wanted more than anything else but men" [16]. Ultimately, though the South always had immense reserves of ore, its underdeveloped iron manufacturing base, chronic labor shortages, woefully inadequate transportation system, and increasing loss of iron-producing territories to the enemy spelled doom in the long, resource-intensive contest against the North.

To oversee the wartime iron industry, the Confederate government charged the Niter and Mining Bureau to monitor production, inspect the factories, and prepare reports for the higher levels of administration. Still, managing officials all too often hindered iron output more than they helped. Concerning Tredegar, for example, Richmond bureaucrats often carped about the firm's work, even hauling a company partner before a congressional hearing in 1863. At the same time, the government frequently did not pay its bills on time or, toward the end of the war, at all. Furthermore, Confederate administrators had promised Tredegar an ample amount of pig iron and coal to meet production quotas, but such assurances for the most part went unfilled. Because of this and its own supply and manpower problems, Tredegar rarely operated above one-third of its full capacity in the four years of conflict.

In spite of all the troubles, Tredegar's contributions to the Rebellion were unsurpassed. Joseph Anderson's business had already made no fewer than 881 artillery pieces before the opening barrage, then put out another 1,160 while the battles raged. One of the most innovative of the heavy guns was the Confederacy's first rail-mounted cannon, an idea suggested by Robert E. Lee in early June 1862. Going to work at once on the general's notion, Tredegar had the weapon ready before the end of the month. This behemoth, a rifled Brooke naval gun that fired 32-pound projectiles, saw action at Savage's Station in the Peninsula Campaign. Brooke guns were named for John M. Brooke, a lieutenant in the Confederate Navy, who had earlier worked with Anderson to design rifled cannons. Four Brookes had gone into battle with the ironclad CSS *Virginia* against the USS *Monitor* in March 1862. Tredegar also armored the Rebel warship, rolling 1.5 million pounds of plate made from pig iron generated by the Catawba Iron Works in western Virginia.

Up to the end of 1862, Tredegar still remained the Confederacy's only supplier of large cannon. Immediately upon completion, many of the big guns were packed off to guard the southern coasts and keep open the gaps in the Union blockade. Late that year another Brooke and Tredegar creation, a seven-inch rifled gun that hurled a 140-pound projectile, bolstered the Confederate arsenal. In April 1863 one of these huge weapons sent a shot smashing through the iron turret of the USS *Keokuk* in Charleston harbor, sinking the Federal monitor almost instantly.

Anderson and company made much more than heavy ordnance in the years of hostilities. Machinery for weapons manufacture shipped out to armories across the South. Tredegar forged the equipment that turned out gunpowder at the gigantic Augusta plant, probably the largest powder mill on either side. The firm served as the core of Confederate ordnance research, participating in experimental work on naval mines (called torpedoes in those days), machine guns, and submarines. Near the end of the war, the patriotic Anderson offered Tredegar's facilities to the government, as he had done at the onset of the conflagration and several times since, but officials turned him down once more.

Fig. 11.1 Ruins of a portion of the Tredegar Iron Works in 1865. Thanks to Joseph Anderson's defense of Tredegar with his men when the Union captured Richmond, most of the physical plant survived intact. Photograph courtesy of the Library of Congress

When Lee abandoned Richmond to Grant on April 2, 1865, civil disorder erupted and rampaging mobs threatened the Tredegar installations. Anderson, a brigadier general who had fought and been wounded in the Peninsula Campaign, led his 300-man Tredegar Battalion in a defense of the plant that saved it from utter destruction (Fig. 11.1). Union soldiers occupied the works briefly during the United States government's attempt to seize Tredegar after Appomattox. Anderson got his company back in September 1865 when President Andrew Johnson, realizing the need to help the South rebuild, pardoned him. The Confederate iron master started up again and Tredegar stayed in business for the next 122 years.

Charcoal Iron Making in Western Virginia

Most of the pig and bar iron that fed Tredegar and the other armaments factories in Virginia during the Civil War originated in the western valleys beyond the Blue Ridge. The two most important production hubs were the Alleghany-Bath District, a cluster of sites in four counties in central-western Virginia, and the Shenandoah Limonite District in the northern Great Valley. Another group of iron works also contributed substantially—the New River-Cripple Creek District in the southwestern counties of Smyth, Wythe, and Pulaski. Of these, the Alleghany-Bath Counties region ranked by far as the foremost source of iron in Virginia; approximately one-half of the total amount of iron ever made in the state came from there.

The primary ore extracted by miners in the western Virginia districts is rich in a mineral called limonite, or brown ore. Limonite occurs when the iron in rocks exposed at the earth's surface combines with oxygen to form iron oxide; in other words, it was basically a kind of natural "rust" being dug out. The limonite deposits are soft, porous, and readily excavated in shallow open pits. The ease of mining and great abundance of the brown ores made them attractive for use in the charcoal furnaces that first appeared in the thick forests of the region in the middle of the eighteenth century.

By the Civil War, charcoal iron furnaces in Virginia had reached their pinnacle of development. The best of these put out a product so good that many iron masters considered it better than the cheaper, mass-produced anthracite iron from Pennsylvania. Choosing a proper site was absolutely essential to the success of the charcoal furnace. The four requirements for profitable pig iron making—seams of ore, wood to make the charcoal fuel, water to power the operations, and limestone for fluxing the ore—had to be plentiful and nearby. In southwestern Virginia, for example, Cripple Creek, a fast moving stream flowing through a densely forested valley floored by limonite-bearing limestone, exemplified an excellent locality for charcoal iron making.

The typical furnace, or stack, had the shape of a truncated stone pyramid (Fig. 11.2). The stack had a square base typically 25 to 30 feet on a side and stood 25 to 35 feet tall. Southern charcoal furnaces employed a technology called "cold blast" in contrast to the coal-fired "hot blast" furnaces more typical in the North.

Fig. 11.2 Speedwell Furnace in the New River-Cripple Creek Iron District. Established in 1799, this is one of the oldest furnaces in the district

Cold blast denotes that a rush of outside air at normal atmospheric temperature was forced or blown into the fire chamber of the stack. In hot blast furnaces, the outside air passed through a series of preheated chambers before entering the fire chamber itself. Flowing water in streams commonly provided the energy to drive the bellows or other blowing devices. Blasting operations began by first loading the stack completely with charcoal and lighting it from the top. Over the next several days laborers known as fillers continuously added mixes of charcoal, limestone, and ore as the temperature gradually went up enough to smelt the ore. The limestone acted to draw off impurities into a slag while the purified molten iron pooled in the bottom of the stack.

In a period of blast, work hands tapped off the iron on the average of twice a day. In this procedure the furnace man wielded a long metal rod to break open a clay plate blocking an outlet at the base of the furnace. Removing the clay door let the hot liquid iron flow into the casting bed on leveled ground in front of the stack. The casting bed consisted of sand sculpted to allow the melt to follow a main trench (the "sow") and then feed into numerous side gutters (the "pigs"), hence the term pig iron. The casting of other items such as pots, pans, skillets, kettles, and stove parts took place using sand molds in a casting shed. A typical furnace could yield five tons of iron daily, consuming about an acre of forest to supply the charcoal for the day's run.

In addition to the basic smelting furnace, many operations included a forge where the blacksmith further refined the pig iron by heating and hammering it into bar iron. Bar iron served as the source metal from which the smithy might fashion wagon wheels, horseshoes, hinges, nails, and various kinds of tools. All told, operating a single furnace and forge combination required over a dozen skilled and unskilled laborers, including fillers, furnace men, casting men, a wheelwright, and a blacksmith.

Charcoal iron making proliferated in the pre-war years in the Great Valley of western Virginia; some 45 new furnaces and forges sprouted up there from 1826 to 1850. One of the most successful endeavors was a furnace and forge in Rockbridge County (so named because of the presence of the famous Natural Bridge) on Buffalo Creek, a tributary of the upper James River in the Alleghany-Bath District. William Weaver owned and operated the works using primarily slave labor. Oddly enough, Weaver was a Northerner and great-grandson of the founder of the Dunkers, a religious sect staunchly opposed to slavery. The Philadelphian arrived in the Rockbridge area in 1823, bought a furnace and forge, and by 1858 owned more slaves than anyone else in the county. Weaver started his enterprise with white workers, but finding them expensive and unreliable turned to slaves, several of whom had been iron smiths back in Africa. Fortunate to have already skilled and hard-working blacks, the iron master treated all of them well and got a faithful, high quality workforce in return.

To begin with, Weaver provided his laborers with good housing in cabins sturdily built of logs and bricks with stone chimneys. He always furnished at least adequate food and clothing, and no record exists of beatings or other harsh measures administered during Weaver's time. He could sell difficult slaves, and did so at times but only rarely, preferring to keep families together as long as they toiled satisfactorily. This unusual owner routinely paid his slaves for work exceeding the production quotas, a practice that caused consternation at a bank in nearby Lexington. A group of the Buffalo Creek black slaves went there to open accounts using their overtime pay. When the bank finally decided not to permit it, Weaver defused the issue by setting up his own savings plan for the workers.

Once the war commenced, Weaver agreed to supply iron to some prominent ordnance-making firms in Richmond, Tredegar among them. Soon the iron from Buffalo Forge and other local works was traveling down the James to Lynchburg and Richmond. Weaver died halfway through the conflict and his nephew Daniel Brady took over the business. A month after peace came, Brady recorded that the former slaves working at the forge were "Declared free by order of the military authorities" [2]. A day later he noted "All hands quit work as they considered themselves free." Shortly after, however, nearly all voluntarily returned to their jobs, a testament to the treatment they received at Weaver's iron works. Buffalo Forge lasted only a few more years, closing with many other Virginia facilities unable to compete with the much cheaper metal coming in from the North and England in the post-war times.

The small iron industry that originated in southwestern Virginia in the late eighteenth century blossomed into the New River-Cripple Creek production district in the nineteenth century. Centered in southern Wythe County, this area rose to dominate iron making in the region until the charcoal furnace era ended after the Civil War. David Graham, the principal iron maker in Wythe County, operated the most elaborate of all the Cripple Creek works. The heart of his empire consisted of a complex of furnaces and forges built along Cedar Run, a feeder stream into the New River. Though several stacks had stood here in the decades prior to the hostilities, only one—Cedar Run Furnace itself—survived into the Civil War period as a significant operation (Fig. 11.3). That furnace, one of several in Wythe County supplying Tredegar, usually brought forth good metal. On at least one occasion, nonetheless, Graham's product proved unreliable when made into cannon. Tredegar informed the ironmaster that several of the guns cast with his iron had misfired or even burst.

The Wartime Iron Industry and Union Raids in Western Virginia

The actual numbers, kinds, and locations of the iron operations in the Confederate states are not known exactly. Poor record keeping by both the iron businesses and the Richmond government is part of the problem, but much information has also just been lost or destroyed. A detailed report from Niter Bureau Chief Isaac St. John dated January 31, 1865, noted that the government operated seven furnaces in the Confederacy in 1864—three in Virginia and two each in Alabama and Texas. Forty-five more furnaces owned by private interests were in blast at one time or another that year, 20 of them in Virginia and nine in Alabama. St. John listed ten of the

Fig. 11.3 David Graham's Cedar Run Furnace in the Cripple Creek area. Built in 1832, Cedar Run provided iron for the Tredegar Works in Richmond. Note the main casting arch at the base of the furnace

Virginia furnaces as "destroyed by enemy," but three had been rebuilt by year's end. He furthermore recorded that 52 forges turned out iron in 1864, 25 each in Virginia and North Carolina.

About 80 per cent of the Virginia furnaces mentioned in the St. John report lay in the western section, concentrated mostly in the Shenandoah and Bath-Alleghany iron districts. By early 1864, Joseph Anderson had gained control of a dozen furnaces in these districts by leasing or outright purchase. He did this to ensure a more reliable supply of high quality pig and bar iron for his Tredegar Works, not at all a certainty considering that the government was redirecting portions of the iron that private companies had always sold to Tredegar to other users. The best raw metal came from Anderson's southern Alleghany-Bath district furnaces, which also had the advantage of proximity to the James River artery to eastern Virginia. Moving his iron by water led Anderson to set up his own boating business, acquiring nine over-sized craft and several smaller ones to carry the heavy loads down the James to the waiting Tredegar mills.

In the New River-Cripple Creek District, wartime blasting occurred intermittently at about half a dozen furnaces. Abijah Thomas owned and ran one of the biggest of these. This entrepreneur built his works on Staley's Creek near Marion in Smyth County and got it into blast just a week and a half before the fighting started. Thomas and his partner W.F. Hurst contracted to sell to Tredegar all of the pig iron made at their stack in the years 1862 and 1863. In a letter dated June 10, 1863,

C.B. Thomas, an agent of the furnace, wrote a contact in Salem, North Carolina, asking the man if he would

> ...do me the favor to get for us 40 or 50 pairs of woolen soles such as you showed me when I was in Salem?... We want them for our hands working about the Furnace and Coaling. They soon burn out leather bottoms [227].

On August 10, 1863, Thomas penned another letter to the same correspondent:

> The Yanks did not extend their visit to our place during the late raid on Wytheville. [This refers to Colonel John Toland's attack on July 18 with about 1,000 mounted men.] I think I could have given them a warmer reception at Marion if they had come on than what they got at Wytheville. We had about 500 men well posted waiting for them [227].

As the war kept going, troubles mounted for the Staley Creek furnace. Obtaining workers became so difficult that Thomas and his son applied to the court on March 22, 1864, for the assignment of workmen, if not over 50 years old, to labor at their stack. The Court responded with 27 men so ordered. On June 22, 1864, and again on November 22, the Court mandated patrols to be active in the section of Smyth County where the furnace was located. These telling events indicated plainly that manpower shortages and increasing Union military pressure now seriously affected operation of the iron works at Marion and others throughout the New River-Cripple Creek district.

Besides the furnaces in Wythe and Smyth Counties yielding raw iron, two foundries contributed significantly to the war effort. Barrett's Foundry in Wytheville manufactured small cannons, cannon balls, and rifles for the Confederate armies. In the summer of 1861, Barrett advertised for 30 gunsmiths to work at the plant where output of rifles had reached ten a day. The business prospered and in 1862, he placed another advertisement for 50 gunsmiths and mechanics. That same year Barrett's shop reworked several hundred old Model 1819 United States breech-loading muskets into more modern muzzle-loading rifles. Conversion of these antiquated firearms involved cutting rifling into the smoothbore barrels and changing the ignition device from flints to percussion caps. Barrett's operation served as one of the few places in the Southern states able to provide percussion caps. From time to time, the Wytheville business performed other armaments work for the national government by repairing damaged muskets and their various parts. Barrett at one point patented an early kind of "gatling gun," devising a weapon with a seven chamber cylinder and operated by a crank.

Partners G.G. Goodell and J.A. Quaife operated another Southwest Virginia foundry providing war-related iron products in Smyth County near Marion. Goodell, commenting on a Saltville contract, wrote in July 1861: "We have to melt 50 tons of iron during the next four months. We have a job of 50 salt kettles that will weigh 1,100 lbs. each..." [6]. One month later, as demand for salt drove higher the need for the kettles and related items, Goodell noted:

> We have plenty of business now, working 21 men and 2 chattels...We are making 5 or 6 kettles a week for which we get 44.00 each...We can not do all the work ordered for want of help. If we had room and machinery we could get work for more than a hundred men [6].

Working at full capacity was not to be, however, for the region's iron works or any of the other mineral industries. Beginning in mid-summer 1863, Union attacks, directed primarily at debilitating the lead mines, salt works, and Virginia and Tennessee Railroad, drastically increased in frequency and intensity. In late 1864 Northern incursions into the area culminated in General George Stoneman's campaign. Although not a major target of the raid, Stoneman's unit found and pillaged several iron operations. On December 16, while in pursuit of Confederates trying to defend Saltville, about 300 Union horse soldiers came unexpectedly upon Thomas' Staley Creek furnace. They razed it almost completely to the ground and destroyed the dam built for water power. A local account of this action noted that the raiders discovered 75 "beeves" (beefs) salted down in a wooden tank onsite at the furnace. The blue-clad troopers took what they could and burned the rest, filling the air for miles with the aroma of roasting meat. On that same day Federal forces reached Wytheville and put Barrett's Foundry to the torch, along with other parts of town.

"Black Dave" Hunter's raid into central Virginia had previously wrecked a number of iron furnaces and forges in summer 1864 (Fig. 11.4). As part of his plan to press Lee from several directions, Grant instructed Hunter to proceed south in the Shenandoah Valley toward Staunton. From there he was to turn east, cross the Blue Ridge, and attack Lynchburg, an important industrial and rail center. Along the way, his men had orders to live off the land and destroy anything in their path that might be useful to the Confederate military.

By early June, Hunter had his riders on the move, crushing a small enemy contingent in a sharp battle at Piedmont on the 5th and occupying Staunton on the 10th. Over the next nine days, Federal soldiers discovered three of the biggest and best Tredegar furnaces—Mount Torry, Cloverdale (Catawba), and Grace—and put them to the torch. In his memoirs, Joseph Anderson described the devastation: The

Fig. 11.4 Union General David "Black Dave" Hunter. Photograph courtesy of the Library of Congress. Instructed by Grant to lay waste to the Shenandoah Valley in summer 1864, Hunter enthusiastically carried out his orders

...South's chief supplier of gun metal, Grace Furnace, another producer of cannon iron, and Mount Torry Furnace...were left in smoldering ruins. Many of the slaves escaped and raiders captured large numbers of draft animals and destroyed extensive stores of provisions [247].

Hunter's foray ended when Confederates under Jubal Early defeated him at the Battle of Lynchburg on June 19, forcing the Union general to withdraw back over the mountains to the sanctuary of West Virginia. Because his three furnaces were fundamental to Southern weapons manufacture, Anderson rebuilt them as quickly as possible. Equipping them with newer and better machinery, he expanded the blast capacity significantly so that by early 1865 these Tredegar stalwarts ran more efficiently than ever.

Raiders like Hunter and Stoneman never seemed to appreciate the real importance of the iron works in their paths, else they might have searched for more of them and done a better job of completely wrecking the ones they found. As it was, western Virginia furnaces kept sending pig iron to Tredegar and other big factories until the very last days of the war.

Notes

The introductory material on Stoneman's raid is from Evans [56]; the Hunter information comes from Walker [248]. The principal source for the discussion of "Confederate Iron and the Tredegar Works" is Bruce [28]. Other references used in this section are Boyle [22], Black [18], Norville [171], Crews [37], Schult [206], and Lynch [127]. For the account of "Charcoal Iron Making in Western Virginia," I took the geology of the iron deposits from Holden [83] and Gooch [69]. Mandigo [128] and Brady [23] provided the general description of the technology of cold-blast charcoal iron making in western Virginia. I used Hoyle [87] for the brief comments on the hot-blast techniques more common in the North. The detailed narrative of the Weaver works in Rockbridge County is from Anderson [2]. The discussion of Graham's furnaces is from Kegley [103]. In the concluding narrative of "The Wartime Iron Industry and Union Raids in Western Virginia," St. John [220] is the report containing the numbers and kinds of furnaces and forges in use for the Confederacy in 1864 that begins this section. The specific depictions of the western Virginia furnaces are from: Whitman [261], Mandigo [128], and Kegley [103] for Wythe County; and Wilson [271], Armstrong [7], and Sturgill [228] for Smyth County. Bruce [28], Ramsey [189], Evans [56], and Walker [246, 248] contributed the information about the Union raids.

Chapter 12
Coal, Confederate Mines, and the CSS *Virginia*

"You cannot do impossibilities"

Sunday, March 9, 1862, ushered in a new era of naval history as two of the strangest warships the world had yet seen clashed on the waters of Hampton Roads, Virginia (Fig. 12.1). That benchmark day witnessed the initial battle between ironclad vessels when the USS *Monitor* and the CSS *Virginia* fought to a draw after pounding each other relentlessly for several hours. One account of the duel noted that "…the Federals were at a loss to know what was propelling the Virginia when it ventured out on the first day of the battle, as they thought that the only coal the Confederates had was bituminous coal which gives off a dense black smoke" [181]. The answer to the Union seamen's befuddlement very likely lay in the kind of coal firing the *Virginia's* boilers. Emitting thick clouds of black smoke, bituminous coal or wood typically did in fact drive most Southern steam engines, but a cleaner burning semi-anthracite may well have powered the Confederate vessel that day.

That specific coal had to have been mined at Merrimac in southwestern Virginia's Valley Coalfields, the only possible source of semi-anthracite in the state or anywhere else in the Confederacy. These coal measures had been discovered in the Great Valley of Montgomery County in the mid-1700s and mines opened up before the century ended. In 1864, a Confederate engineer drew a map of Montgomery County that showed the location of a "Govt Colliery," or government coal mine, at Price Mountain near Blacksburg. Colliery, a British term, was then widely used in America to refer to coal diggings; accordingly, the map confirms the presence of an active mine in the southwestern semi-anthracite belt during the Civil War. The distinctive Price Mountain coal is likely the fuel that was shipped to Norfolk and loaded into the bunkers of the mighty *Virginia*.

The Montgomery County mine production was important, but the great preponderance of the coal contributed by Virginia to the Confederate war industries came from the eastern Piedmont region. A huge geologic feature called the Richmond Mesozoic Basin formed there in the Age of Dinosaurs and filled with sedimentary

© Springer International Publishing Switzerland 2015
R.C. Whisonant, *Arming the Confederacy*, DOI 10.1007/978-3-319-14508-2_12

Fig. 12.1   USS *Monitor* (left) and CSS *Virginia* engage in the world's first battle between ironclad warships. The Southern vessel was the former USS *Merrimac*, a scuttled wooden steam frigate raised and plated with Tredegar armor by Confederates. She was renamed the CSS *Virginia* before sailing into battle. Image courtesy of The Library of Virginia

layers containing compacted and combustible plant remains—coal. Extensive and relatively easy-to-dig bituminous seams crop out a few miles west of Richmond where local people first found and used them around 1700. Known generally as the Midlothian mines, numerous shafts and pits had been worked by the 1860s. The Midlothian product held enormous importance in the conflict in that it provided the main energy source for Virginia ordnance manufacturing.

These same mines also served as the only significant supply of coal anywhere within the Confederacy. During the Civil War, Richmond coal shipped out to several other states to drive steam engines in everything from armored naval vessels to heavy industry factories. Recognizing the strategic importance of the Midlothian fields, Tredegar owner Joseph Anderson led the fight to keep them going. Anderson worked tirelessly to replace and maintain the mining machinery and to procure the hard-to-come-by labor to do the work. For the most part, his efforts succeeded and Richmond coal helped support the Rebellion until the surrender at Appomattox.

Northern vs. Southern Coal and Iron Production

When civil war came, a major geological difference existed between the coal resources of the Northern and Southern states, one that had a profound impact on the growth of heavy industry in the antebellum years. The difference was that massive deposits of anthracite, or hard coal, resided in Pennsylvania whereas practically none occurred in the South. Anthracite is basically a metamorphic rock, meaning that elevated heat and pressure generated by crustal movements have driven out most of the impurities found in lower grade soft coals. The end result is a kind of natural "coking" process that creates a hot-burning, carbon-rich fuel extremely desirable for the high-temperature furnaces making iron and steel. Furthermore, steam engines in military applications such as train locomotives and naval vessels run much more efficiently with anthracite driving them.

Turning accumulations of vegetation into coal can go through four successive stages: peat (the lowest energy form), followed by lignite, then bituminous coal, and in rare cases anthracite (the highest energy product). Making anthracite is a relatively uncommon geologic event given that the coal must be held in just the right high temperature and pressure range to force out the non-carbon components but not destroy the coal completely. Bituminous coal is the usual final step in the coal-forming process and is found in many places around the world. Anthracite is much less common and in the United States is confined almost exclusively to Pennsylvania.

In the mid-nineteenth century, northeastern Pennsylvania possessed three-fourths of the world's known anthracite deposits, a geologic gift that had made possible the rise of the "hot blast" method of making pig iron. This metal proved much less expensive to make than that from less efficient charcoal-burning furnaces, yet had a more than satisfactory quality. Copious amounts of anthracite prompted a fundamental relocation of American iron manufacturing to Pennsylvania where the operations grew rapidly as the young nation industrialized. Supporting enterprises such as railroads and massive port facilities at New York Harbor and Philadelphia emerged to handle the escalating iron and steel output. When open warfare finally erupted in 1861, five anthracite railroads operated in the North and the region's heavy industry was well on its way to world-class status.

In contrast, the states that would comprise the Confederacy, with the notable exception of Virginia, failed to create strongly linked coal and iron industries. Bituminous coal had been found in many locales, most notably at Richmond and in western Virginia, eastern Kentucky and Tennessee, and central Alabama. Even so, wood remained the fuel of choice—it was more abundant; closer at hand to most manufacturing, particularly the pig iron furnaces; and generally cheaper to use. Most southern locomotives, for example, still burned wood in the war years. In Virginia, however, coal was in high demand for the large forges and foundries in Richmond, thanks primarily to the already developed bituminous deposits in the Midlothian fields a few miles away. Before long, a small capital city railroad network, similar to the much bigger Pennsylvania system, had grown up to haul the Midlothian coal to the waiting iron manufactories or shipping docks on the James.

In western Virginia and the rest of the Southern states, iron masters preferred the charcoal-fired "cold blast" technology to make the raw pig and bar iron that went to the finishing shops and mills. In fact, as mentioned previously, many considered the quality of charcoal iron to be better than the mass-produced Pennsylvania metal, but charcoal furnaces had reduced capacities and were much less efficient. By comparison, the anthracite-fueled furnace claimed at least twice the capacity and a much faster rate of yield, making the cheaper and more plentiful Pennsylvania metal products very difficult competition for Southern producers. Ultimately, the geological inequity of the world's most extensive anthracite reserves being located in the northeastern United States led directly to making that section a pre-war heavy manufacturing superpower. Coal, iron, and railroads were the ruling triumvirate of the new industrial warfare emerging in the 1860s, and the North had them all in abundance. The South did not.

Richmond Basin Coal

The Midlothian mines occur in the Richmond Mesozoic Basin, one of a series of fault structures that ripped open in the present eastern seaboard states about 200 million years ago. At that time, the Pangean supercontinent began breaking asunder, and along the eastern verge of what would later become North America, rift valleys appeared. Stretching of the earth's crust when continents pull apart causes tremendous masses of rocks to drop down along faults, creating immense valleys on the earth's surface. Today, the Olduvai Gorge in East Africa and California's Death Valley are in rift-generated depressions.

When the Richmond Basin formed, generally warm and wet climatic conditions prevailed, and lakes, swamps, and bogs dotted the valley floor. Early smaller versions of the later Mesozoic giant dinosaurs foraged among the thick stands of vegetation covering the marshy lowlands, leaving their footprints by the thousands in the soft mud bottoms. Eventually Pangea fractured completely apart; the rift basins grew inactive and in the end filled up with layers of sediments. Burial beneath the thick overlying strata compressed and altered the ancient masses of plant material into bituminous coal. Subsequent uplift and erosion have once again brought the carbon-rich beds to the earth's surface in a series of disconnected Mesozoic Basins extending from Connecticut to Georgia.

French Huguenot settlers discovered coal in the Midlothian area near Richmond perhaps as early as 1699; in 1704 Colonel William Byrd obtained a patent for 344 acres of the coal-bearing lands. Other local entrepreneurs claimed properties in the vicinity over the next several decades with the "...liberty to dig coal" [79]. Commercial-scale production of coal initiated in America in 1748 in the Midlothian fields. The earliest shipments in the country took place two years later when loads of this coal traveled from the port of Richmond to Philadelphia. (The Virginia capital could be reached by ocean-going vessels sailing up the broad estuarine James River, thus docking and loading facilities for maritime traders had long been present there.) Transport to New York and Boston followed shortly thereafter.

By the early nineteenth century an especially productive Midlothian operation, the Black Heath pits, had achieved considerable fame in coal mining circles. An American Journal of Science article in 1818 described the mines as now having

...three shafts open. A steam engine, constructed by Bolton and Watt – is used exclusively for pumping out water. The coal is raised in a box, called by the miners cowe (corve). These cowes contain about two bushels each, and two of them are alternately rising and descending in each shaft. They are raised by means of ropes, fastened to a simple wheel and crank, which is turned by mules [252].

In the Civil War high quality Black Heath coal would be used exclusively in Richmond's cannon foundries where iron smelted from western Virginia ore was cast into heavy guns for the military.

As the Midlothian works expanded in the antebellum years, the technology of coal extraction improved. The earliest users of the coal simply dug it from pits along the outcrop trend, but soon shafts had to be sunk to follow the seams dipping farther underground. Shaft mining came to predominate by the mid-eighteenth century, some of the shafts descending as deep as 775 feet by the 1840s. The bigger mines at this time had as many as 150 men and boys and 25 mules working night and day, six days a week with Sunday off to haul out the water. Mules drove much of the equipment, the animals having been lowered on ropes underground where they were fed and stabled. Despite not seeing the sunlight, the mules were maintained in good health, one observer writing in an 1846 newspaper article that "Nevertheless they [the mules] keep fat" [253]. That same newspaper ran an advertisement seeking Negro men to work as miners, noting that

> A well conducted Hospital, under the care of a careful steward and daily attended by three Physicians, is provided at the Mines. – The slaves have a Church of their own at the Pits, and divine service is regularly performed on the Sabbath by white Ministers [253].

After 1830 workmen used black powder to blast the coal loose underground. It was then shoveled into carts and carried over to the shafts where mules or steam engines hoisted the shattered blocks to the surface.

A noteworthy advancement in technology appeared in 1831 when Colonel Claudius Crozet built the first railroad in the state to transport the Midlothian coal away from the mining sites. The innovative Crozet, a French engineer who once served as an artillery officer in Napoleon's army, had immigrated to Virginia where he became the state's original Chief Transportation Engineer. His new Chesterfield Railroad carried coal 13 miles from the Black Heath mines to shipping facilities on the James River at Manchester (Fig. 12.2). The loaded cars moved by gravity from the mines over steel-capped wooden rails downhill to Manchester. Mules rode in the last car, then pulled the empty containers back uphill to the excavations. Steam locomotives followed in the 1840s, causing the rail lines to proliferate dramatically. By 1858, a network of half a dozen railroads crisscrossed the Midlothian mining locality, sections of which survived the hostilities and exist today as part of the Norfolk Southern or CSX systems.

The coal beds in the Richmond Basin emitted dangerous amounts of methane gas which the miners tried to remove by holding lit candles to the coal face using pulleys from a distance. Although this method of burning the gas off held much risk for the laborers, it commonly did make the area to be worked safer. Ventilation was poor, however, and explosions frequent, as for example the disastrous fire that killed 53 of 56 miners at Black Heath in 1839. Regardless of such hazards, the work went on and the mines prospered. In 1850 the Midlothian measures had 14 active digs that turned out about 200,000 tons of coal that year. An 1858 map depicts 47 open pits and shafts that operated at one time or another in these coal fields.

When the Civil War began, Midlothian mining activity stepped up, but shortly thereafter, output started dropping off, declining to only 40,000 tons in 1863. Much of

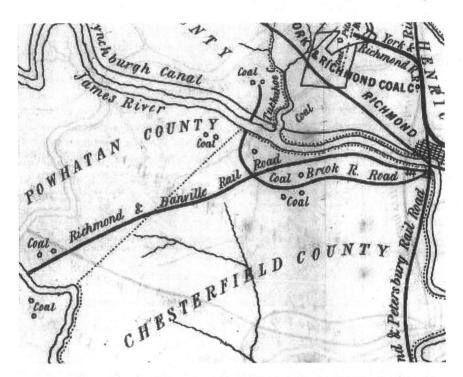

Fig. 12.2 1856 map showing the Chesterfield Railroad (labeled as the "Coal Brook R. Road") at Richmond. The first railroad in Virginia, its cars moved by gravity and mules. The line carried Midlothian coal from 1831 to 1851 when it was replaced by a steam-powered railroad. Image courtesy of the Library of Congress

the wartime coal was prioritized for use in the Tredegar works to keep the weapons-making giant running, though even there the fires went out for a few weeks in the winter of 1862–1863 owing to inadequate fuel delivery. Frustrated owner Joseph Anderson wrote to one of his coal suppliers in early 1863 that "The heavy operations here have been suspended for want of coal since the 23rd of December… and since that time we have not been able to cast a single gun…You cannot do impossibilities" [27].

Foreseeing such problems earlier, Anderson had been buying or leasing formerly active coal digs to keep his furnaces and foundries going, furnishing the old mines with new equipment manufactured at his own plant. He hired extra overseers for the coal works and improved transportation of the coal from mine to factory. Persistent manpower shortages presented a more difficult problem to overcome, yet Anderson managed to plead successfully to keep some of his slave miners exempt from government impressment. Even if there been a full labor force, Tredegar would not have operated during the war at full capacity because the necessary 20,000 tons of raw pig iron and 100,000 tons of coal to do so just could not be produced.

Still, thanks to Anderson's administrative skills, save for that one brief interruption of work "for want of coal" early in the fighting, the firm's books never recorded another wartime shut down due to lack of fuel deliveries from Midlothian mines. The combination of Richmond Basin coal and western Virginia iron ore made metal that rained destruction on enemy soldiers and sailors for the rest of the conflict.

Montgomery County Coal

Coal mining in western Virginia started up about half a century later than in the Midlothian fields. Settlers spreading west across the Piedmont and over the Blue Ridge happened upon coal outcrops in the Great Valley of Southwest Virginia sometime around the mid-eighteenth century. These seams varied significantly from the Richmond Basin bituminous layers—the black fuel was harder, hence the term "stone coal," and burned hotter, a heritage of its very different geologic history.

The Valley coals lie in a belt running through Montgomery and Pulaski Counties and are particularly plentiful in the Price Mountain area near Blacksburg. This locale is in the heart of the Valley and Ridge Province where the ancient sedimentary beds have been intensely folded and faulted. The deformation of the rocks occurred when they were crushed between colliding continents as Pangea assembled around 250 million years ago. A hundred million years earlier in the Mississippian Period, the Appalachians had just commenced to uplift. Along the western margin of the young mountains, masses of mud, sand, and gravel poured off the rising peaks, building vast coastal plains and deltas into a shallow sea that covered most of the rest of proto-North America. On this swampy apron of sediment, jungle-like forests, home to early amphibians hunting their insect prey, thrived in the warm climate and produced thick deposits of vegetation. Buried deeply, then further refined by the heat and pressure exerted in the continental collision zone, the plant matter transformed beyond bituminous grade into metamorphosed semi-anthracite coal.

Unfortunately for the men who would one day work in the mines, less desirable changes accompanied the enhanced energy content of the stone coal. Coals metamorphosed to this degree typically generate perilous amounts of methane gas and display a great deal of fracturing. The large quantities of methane, which may yet have value as a future fuel resource, created dangerously gassy shafts that sometimes exploded violently over the long history of mining in this region. Moreover, the extensive breakage of the rocks produced inherently unstable mine walls and roofs, resulting in the all too common cave-ins that claimed more lives. In spite of these menacing geologic conditions, many attempts have been made to extract the coal, especially in the Price Mountain area. Historic maps show 181 mines and prospects active at one time or another in that location alone.

The first reports of coal discovery in what would become the Valley Coalfields are from Montgomery County around 1750; simple open pit mining appears to have been underway 20 years later. Some accounts of the earliest days of the Valley coal industry contend that foreign mercenary soldiers played a prominent part in working the seams. During or shortly after the Revolutionary War, men from a regiment of Hessian troops taken prisoners were brought to southwestern Virginia to avoid recapture by the British. A number of these German natives who had been coal miners and iron masters in their homeland found employment putting those skills to use digging in the local pits. A slight variation of this story says that the foreign captives arrived voluntarily in Montgomery County with Colonel William Preston after the war and stayed to build their homes and livelihoods in the region. By the 1790s a

Hessian named Jacob Broce seems to have been operating a coal mine, iron smelting works, and a rifle barrel boring mill along a creek in the Price Mountain vicinity.

Throughout the late eighteenth and early nineteenth centuries, small-scale mining of the stone coal took place at Price Mountain and in other parts of Montgomery County where the beds cropped out. Most of this production went to nearby iron smelters, blacksmiths, and farmers. These earliest mines were surface diggings, but eventually coal extraction shifted to underground drift and slope mines. A drift mine is a shaft dug on a horizontal level, whereas a slope mine follows the inclination of a coal seam as it plunges downward at a dipping angle. Slope mines grew to dominate mining throughout the Valley fields because the tectonic stresses put on the rocks by mountain building left the coal beds slanted in various directions. Output in the antebellum years remained modest; the 1840 United States minerals census reported only 200 tons of coal production in Montgomery County.

The remoteness of the Valley coal fields greatly hindered the growth of the industry in the first half of the nineteenth century. No railroads existed in Southwest Virginia, thus horse-drawn wagons, a very costly means of transportation, carried the coal to the James River. It then moved by water to eastern Virginia. Interest in the hard coal seams soared in the 1850s, however, when the tracks of the Virginia and Tennessee Railroad at last extended through far southwestern Virginia. That same decade saw four major companies arise in Montgomery County, the most notable being the Price's Mountain Coal Mine Company. This corporation apparently survived into the Civil War years and may very well have worked the mine from which the Confederate government obtained coal.

Three partners—Jeremiah Kyle, W.M. Montague, and W.D. Kyle—incorporated the Price's Mountain Coal Company in April 1853. The enterprise came to be known as "Kyle and Montague" and this is its listing in the United States Industrial Census of 1860. In that year, the company output dwarfed all other Valley fields operations, putting out 2,857 tons of coal valued at $11,200. No other mine extracted coal worth more than $500. In 1927 Mr. Guy Ellett, a Montgomery County resident, penned a personal letter to Mrs. J.C. Peter in Radford describing the early history of the various mining endeavors at Price Mountain, including those of Kyle and Montague:

> Prior to the Civil War [the coal] had been used for blacksmith and smelting purposes to a limited extent. Mr. Jeremiah Kyle, who was a very prominent merchant residing in Christiansburg became interested in the coal and brought some Welsh miners from Wales who drove three slopes at different points on the outcrop of the coal for a distance of about 400 ft. These slopes were put down for the purpose of proving that the coal was regular and it was not a pocket as was asserted by some geologists at that period. After these slopes were driven, a Company was formed for the purpose of building a railroad from the mine to Christiansburg in order to have transportation facilities [182].

That railroad was not built, and all work halted at Kyle and Montague when armed conflict broke out in 1861. Nevertheless, important times lay ahead for Price Mountain coal, times that included wartime mining for the Confederacy.

Did Price Mountain Coal Power the CSS *Virginia*?

The naval action between the USS *Monitor* and CSS *Virginia* was not only the first fight between iron-plated warships but also the first in which coal fired both vessels entirely. The story that Price Mountain coal supplied the fuel for the original Rebel armored gunboat is deeply ingrained in southwestern Virginia lore. As noted earlier, the existence of a coal mine in that locale operated by or for the Confederacy is substantiated by an 1864 map drafted by a Confederate engineer showing a "Govt Colliery" (Fig. 12.3). The location is about a mile from the present-day community of Merrimac, a small mining settlement named in the post-war years for the sunken wooden Union ship that was raised and converted into the Southern ironclad. (Before the former USS *Merrimac* actually sailed into battle with the USS *Monitor*, Confederates rechristened her the CSS *Virginia*.) The colliery could very well be the Kyle and Montague mine, but no official records have been located that specifically identify the owners. Nor has documentation emerged that indicates where this coal went or how it was used. Despite the lack of firm evidence, the tradition developed after the war that Price Mountain coal did in fact drive the engines of the *Virginia*.

A mine belonging to Kyle and Montague is known to have operated at Price Mountain in the Civil War years, although activity evidently stopped for a time at

Fig. 12.3 Portion of the 1864 Confederate government map of Montgomery County. Note the location of a "Govt Colliery" (center right of the map). According to some, coal from this mine fueled the CSS *Virginia* in her battle with the USS *Monitor*. Image courtesy of the Library of Congress

the outbreak of hostilities in 1861. Company partner William Montague recalled in 1894 that his mine turned out coal for the forging of salt kettles and the making of shot and shell for the Confederate army. Virginia Polytechnic Institute mining engineering professor O.C. Burkhart, an authority on the history of the Price Mountain mines near Merrimac, stated in 1931 that a "…fair sized camp was built [in the Merrimac vicinity], and…the Confederate government mined coal for use in the southern war area" [29]. Burkhart went on to say that "Coal from this mine supplied the *Merrimac*…and it was from it that the mining camp and the coal seam were named" [29].

A version of the *Monitor* and *Virginia* encounter appeared in a 1930 magazine article based on the observations of a survivor of the historic battle. The writer noted that the Confederate government took over a Price Mountain mine of unspecified ownership once the war started and extracted sizeable quantities of coal. The article also claims that this coal eventually was shipped down the James River to Norfolk "…to furnish steam for the ironclad" [177].

The 1927 letter cited above by Guy Ellett, whose father served as a Confederate colonel and medical doctor, provides detailed information related to Montgomery County coal and the CSS *Virginia*. Ellett wrote:

> About this time the Civil War broke out and all work was suspended. The only coal at the time of the outbreak of the Civil War that was being operated in the South was at the coal basin near Richmond which is situated in Chesterfield and Powhatan Counties in this State. After the Federals occupied that territory it was impossible to mine any of that coal for manufacturing purposes and after the Merrimac, which had been sunk in Norfolk harbor and the Navy yard was destroyed, it was raised and re-modeled and renamed the Virginia, there was then the problem of securing fuel to feed the boilers of this vessel and this problem was up to the officers of the Confederate Navy and they had heard of the coal deposit in Montgomery County and upon investigation they employed Mr. I.H. Adams, now deceased, but formerly head of Adams & Payne in Lynchburg and of Adams, Payne & Gleaves in Roanoke, to manage the coal operation which would supply the coal necessary to feed the boilers of the man-of-war Virginia. Mr. Adams told the writer that he and his family moved to the mine and at that time there were some 40–50 houses with a very good mining plant with bunkers and mechanical hoists on the property which had been placed there by the Welsh miners when they were proving the coal seams; that he filled these houses with miners and had quite a large production per day at that time; that for transportation they hauled the coal by six mule teams to Christiansburg where it was shipped in cars to Norfolk and placed in the bunkers of the man-of-war. I think if you will refer to a book written by Mary Johnston, I do not recall the name, she mentions this fact in her book and states that the coal that was burned on the Virginia came from far-away Montgomery County [183].

What became of the Price Mountain coal operations as the Civil War drew to a close is unclear. Some sources indicate that Union soldiers under General William Averell raided the coal facilities in 1864 and destroyed the mines. Others claim that General George Stoneman's troopers demolished the coal works. Both commanders led forces through Montgomery County late in the war, yet neither mentions the wrecking of any coal mines in their after action reports. Indeed, no Union or Confederate official accounts or any contemporaneous newspaper articles speak of a Price Mountain wartime mine or the destruction of the facilities there.

Montague himself made no mention of Federal raids in his 1894 remembrances, but then he made no reference to his coal fueling the *Virginia* either. Perhaps, like many other Southern industrial activities near the end of combat, work at the Price Mountain coal mines diminished or ceased altogether as manpower shortages became acute and war fatigue set in.

Four decades after the fighting stopped, large-scale workings began in 1902 and the central mining location at Price Mountain was given the name "Merrimac." At this time, the Virginia Anthracite and Coal Company, owner of the operations, widely circulated the story that Merrimac coal had stoked the engines of the Confederate ironclad. Did Montgomery County stone coal provide the fuel for the South's first armored warship? The question remains unanswered.

Notes

The beginning paragraphs concerning Montgomery County coal and the CSS *Virginia* are from Pierce [177] and Proco [184]; the following overview of the significance of the Midlothian mines to the Confederate war effort is summarized from Bruce [28], Weaver [254], and Hibbard [81]. The next section comparing "Northern vs. Southern Coal and Iron Production" is condensed from two sources: Hoyle [87] (Northern) and Bruce [28] (Southern). For the discussion of "Richmond Basin Coal", the geologic history that begins the section is my summary of information found in standard geology texts; again, Hamblin and Christiansen [75] and Tarbuck and Lutgens [231] are very good. The history of mining activities in the Richmond coal fields I found in Weaver [254], Wilkes [267], and Hibbard [80]. Bruce [28] is the source of the information on the strong connection between Joseph Anderson and his Tredegar Works with the Midlothian mines. For the description of "Montgomery County Coal," I used Burkhart [29], Worsham [278], Hibbard [80], Bartholomew and Brown [13], Price [180], Proco [184], Freis [66, 67], and La Lone [116]. (The geologic history that begins this section is mine, again based on current geologic thinking.) The concluding section questioning whether or not Price Mountain coal actually powered the Rebel ironclad is grounded in Proco [184]; Pierce [177], Burkhart [29], Worsham [278], and Freis [67] also contributed.

Chapter 13
Confederate Railroads

"The gut of the Confederacy"

The celebrations began long before workmen pounded in the last spike to complete the Virginia and Tennessee Railroad through southwestern Virginia in October 1856. Two years earlier, when the line reached Central Depot—so named because it lay halfway between the two terminal points at Lynchburg and Bristol—crowds drawn by the offer of free barbecue flocked to the festivities marking the event. A special round trip excursion traveled over the advancing rails from Lynchburg, where one local business attempted to profit from the opening of the railroad at Central with this advertisement: "Ho! For the New River! The Great Excursion on the 1st of June/Persons going to the New River on the 1st of June can be supplied with a tasty outfit by calling at the Franklin Cloth and Clothing House" [277]. The Depot's new rail facilities included freight buildings, a machine shop, and an engine round house. Most impressive was the long bridge across the New River still under construction less than two miles away (Figs. 13.1 and 13.2).

No revelers on that happy June day in 1854 could have foreseen that ten years later the span and the town itself would be blasted by a torrent of shot and shell from artillery belonging to the United States Army. During the Civil War, the Virginia and Tennessee would turn into one of the most valuable stretches of track in the Confederacy, hauling enormous amounts of strategic materials, including mineral products, timber, and foodstuffs to sustain the fighting. And the North would become determined to cut that vital lifeline by destroying its most crucial structure—the bridge over the New River.

© Springer International Publishing Switzerland 2015
R.C. Whisonant, *Arming the Confederacy*, DOI 10.1007/978-3-319-14508-2_13

Fig. 13.1 1856 sketch by artist Lewis Miller showing the Virginia and Tennessee Railroad Bridge over the New River at Central Depot. Later on, workmen added a wood-sided, tin-roofed covering to complete the structure. Image courtesy of the Virginia Historical Society

The Iron Horse: A New Factor in Warfare

The American Civil War was a war of "firsts." Startling new weapons and technologies were either invented or initially used for military purposes during the conflict, including the telegraph, aerial observation, iron ships, submarines, and machine guns.

Fig. 13.2 The Virginia and Tennessee locomotive "Roanoke" in 1856 in Abingdon, Virginia. The "Roanoke" was one of the first engines on the line and operated out of Central Depot for many years. Photograph from Norfolk and Western, Colonel Lewis Jeffries Collection

Yet no recent innovation had more impact than railroads. The idea of moving wheeled vehicles along tracks to haul freight or people goes back to the ancient Greeks who cut grooves in limestone along which animals pulled wooden carts. Wooden-railed track ways appeared in Germany in the sixteenth century, and in the 1790s iron began to replace the wood in English track; however, the motive power remained animals. That changed dramatically in 1812 when two British inventors developed the first commercially successful steam-driven locomotive. By 1825 British innovation had led to the world's first public railroad that hauled freight (and soon passengers) using steam locomotives. Railroads continued to spread across the countryside in Great Britain over the next decade and by 1840 a more-or-less integrated network had taken shape. In 1855 the first military use of railroads occurred when the British Army built a seven-mile-long track to supply troops besieging Sevastopol in the Crimean War.

Railroads reached the United States in the early 1830s whereupon construction of this new marvel soared. The integration of railroads into the young nation's transportation system in the following years was nothing short of astonishing. McPherson pointed out that "The 9,000 miles of rail in the United States by 1850 led the world, but paled in comparison with the 21,000 additional miles laid over the next decade, which gave the United States a bigger rail network than in the rest of the world combined" [145]. The results of this stunning increase were extraordinary. Older settlements on the main lines underwent explosive growth and in many places brand new cities sprang up along the rails. An undistinguished town in the continental interior named Chicago became an important rail hub and mushroomed into a major metropolis. The iron horse spurred economic development, particularly in manufacturing and heavy industry. Feeder industries, notably iron and coal, surged.

When the Civil War broke out in 1861, the iron railways quickly came under assault by the newly assembled armies. One month after Fort Sumter, an obscure Confederate colonel in western Virginia named Thomas J. Jackson seized control of 44 miles of track on the Baltimore and Ohio (B&O) in the northern Shenandoah Valley. Not long after, he confiscated several dozen B&O locomotives and 300 freight cars, destroying many while sending some to Virginia railways for use against the enemy. Later in 1861 Jackson ripped up about 20 miles of rails on the B&O west of Harpers Ferry and shipped them south to other routes desperately in need of replacements.

The difference the iron horse could make in a big battle became obvious early in the conflict. In July 1861, 35,000 eager, but inexperienced, Union troops under General Irvin McDowell left Washington to attack the Confederate rail center at nearby Manassas, Virginia. Some 11 railroad companies ran into this hub where 11,000 equally untried defenders commanded by General P. G. T. Beauregard had been hastily gathered, many brought there by the Manassas Gap Railroad. Just before the shooting started, Beauregard's outmanned forces were swelled significantly by fresh units under Generals Joseph E. Johnston and Thomas J. Jackson rushed in over the same tracks. The concentrated Southerners gained the victory, signaling clearly that steam-powered engines could affect battles decisively by moving large numbers of soldiers to fighting fronts far faster than they could have marched.

Beyond the tactical value of railroads in transporting men and their armaments to the combat zones, broader uses to carry foodstuffs and raw materials soon became essential for the total war effort. Throughout the struggle, attempts by both sides to preserve their own system and strike at the enemy's network intensified. When the railroads supplying Atlanta were lost in 1864, that city fell. After Grant cut the rail lines feeding into Petersburg in 1865, Richmond had to be abandoned. The war effectively ended seven days later at Appomattox.

Deficiencies of the Rebel Railroads

In the years before the war, expansion of American railroads and the flourishing manufacturing and industrial centers spawned by them had not proceeded at the same pace in all sections of the country. By 1861 the Confederacy had 9,000 miles of track, more than most nations. Still, that was dwarfed by the 22,000 miles of rail arteries constructed in the northeastern states and the free states in the upper Mississippi River Valley. Moreover, the North claimed nearly all of the factories for making rails, locomotives, and cars. Of 470 locomotives built in the United States in 1860, only 19 originated from shops in the South. During the war years, Pennsylvania alone would produce twice as many railroad cars as the entire Confederacy. The Confederate states failed to turn out a single new rail after 1861.

This great discrepancy between Northern and Southern rail advancement, as well as industrialization in general, stemmed from a number of causes. Slaves and cotton, the major sources of Southern wealth by far, were in fact the prime disincentives to develop transportation, manufacturing, and infrastructure. Cheap slave labor made possible the hugely profitable large cotton plantations that required very little modern farm equipment (and the factories to make it), just more slaves and more land to operate and expand. Rails to move the cotton were not needed; the numerous rivers in the South transported the crop to market more than satisfactorily. Another factor holding back modernizing the Southern economy was the self-interested profit-seeking of the large estate barons who did not want industry and infrastructure competing for resources. Finally, a widespread cultural disdain existed in the South for Yankee industriousness and pursuit of profit from mechanized businesses. In the antebellum years, many plantation masters looked derisively at the North's racing ahead to establish a world-class transportation and manufacturing base, deeming such activities as unworthy of an elite agrarian society.

And yet, in those decades before the 1860s, the South had made considerable progress in railroad construction. Three primary centers grew up at Chattanooga, Atlanta, and Richmond, and a host of lesser cities came to be rail heads as well. These interior lines proved to be an enormous advantage during the war. With less distance to travel, troops and supplies moved over the steel rails swiftly and sometimes decisively to battles, as at First Manassas in 1861 and Chickamauga in 1863. From the very beginning, nonetheless, several weaknesses in the Confederate rail service kept it from ever being fully utilized. One of the worst deficiencies was that the system never achieved true integration. Over 100 companies comprised the Southern complex, but their lines tended to be short and often did not physically connect. The average length measured about 85 miles, and only nine routes reached more than 200. The Virginia and Tennessee Railroad stood out as one of the longest, running 213 miles through the resource-rich region of Southwest Virginia.

The lack of connectivity presented a serious problem. It meant that at transfer points, freight and passengers had to be taken off one train, conveyed by foot or wagon over to the next carrier, and then loaded into the new cars. As early as June 1861 Robert E. Lee sought to get at this difficulty by having more connecting tracks built in the Richmond-Petersburg area. Many of the railroad operators themselves favored this and other attempts to better link up the lines, but local authorities resisted out of concern for lost revenue. Servicemen and other riders would no longer seek accommodations while they waited for another train to keep going on their journey, nor would businesses be needed to haul the cargo.

Equally vexing at nodes of transfer was the variety of track gauges, ranging anywhere from four to six feet. For example, lines in Virginia and North Carolina, through which nearly all of the eastern traffic had to move, tended to be standard gauge, whereas most of the other states had broader gauges. Finally, the railroad firms, although not opposed to better connections and uniform gauges, did not like to share their tracks or cars with competitors and definitely did not want the national government interfering in the running of their businesses. In the end, despite attempts by Confederate officials to improve things, the gaps and gauges shortcomings proved intractable. Delays in moving men and equipment effectively created by these obstacles only worsened as the war wore on.

The Richmond government did take measures to alleviate the difficulties besetting the rail network. Meetings with rail executives convened in April 1861 to set standard charges for carrying troops and military supplies. That summer Confederate President Jefferson Davis appointed a former railroad president to administer military rail activity in Virginia. Beginning in late 1862, the first of a series of overseers of all the Southern lines took office, yet none proved effective in the long run. Labor difficulties surfaced early, in large part because many of the rail workers came from the North and, when war began, returned home to toil for the Union. The government Conscription Act of 1862 showed consideration for the rail workers by exempting several thousand, only to have new legislation in 1864 take away some of those generous provisions. Slaves made up a significant component of the railroad labor force throughout the hostilities, usually in the more menial tasks. As was true for all Confederate industry and military services, however, there never was enough manpower to keep the trains running in the long grinding war the South found itself fighting.

With the onset of heavy wartime usage, the inadequacy of the Southern rail system rapidly became evident. Half way through the conflict, the task of simply keeping the trains running threatened to overwhelm Confederate government and military officials. The tracks, locomotives, and cars wore out fast and the South lacked the heavy industry necessary to replace the disintegrating equipment. By fall 1862 quarrels between administrative agencies scrambling for the pitifully thin available stores of railroad iron had morphed into a serious issue for the Confederacy. Forced into regulatory action in January 1863, the government in Richmond formed an Iron Commission to supervise the reallocation of the metal from the smaller routes to others considered more essential.

A crisis of a different sort emerged by 1863 with a lack of wood for fuel and replacement of cross ties and bridge timbers. An absence of workers to cut the wood brought about this problem. On occasion, passengers, especially soldiers, had to disembark and chop timber to fire the locomotive boilers so the trip could continue. (Unlike Northern locomotives that burned coal, Southern ones were fired by wood.) Ties as well as the rails themselves more and more were cannibalized from less critical lines to keep the main arteries operating.

As the tide of battle shifted in favor of the North, its advancing armies captured an ever-increasing amount of Confederate territory, resulting in more railroads either being destroyed or rebuilt for Union use. Sherman's summer 1864 Atlanta Campaign showed this general's recognition of the value of the steel highways. Before he left Chattanooga, Sherman trained about 10,000 troops in railroad construction and repair. Following the Western and Atlantic Railroad south, his soldiers tore up the tracks and ruined the rails by heating them in blazing bonfires of ties until they glowed red hot. They then bent the iron around trees to make the famous "Sherman's Neckties." After Atlanta fell, Sherman abandoned his own rail supply links and commenced the March to the Sea. His trained repair gangs had the skills to speedily restore the lines wrecked by the retreating enemy, often on the same day.

Meanwhile, destruction of tracks useless to him went on, even more so in South Carolina, the starting point and heart of the Rebellion.

Ironically, perhaps the most basic reason for the inability of the Confederate railroads to properly support the war was ideological—the principle of States' Rights. Most carriers operated within a single state where owners and government officials alike clung to their opposition to national coordination from Richmond. In the North, Congress gave President Lincoln the power to take over the railroads for military purposes and established the United States Military Railroads as a separate government agency in 1862. In contrast, President Davis's administration never set up the proper government supervision needed to maximize efficient use of the rails and never provided the centralized maintenance or protection required to keep the trains going. Rather, the companies within the states ran their rail lines as separate fiefdoms, resistant to being part of a national system to the end. To allow central control would have been to abandon the most fundamental doctrine that the Confederate states struggled so mightily to defend.

Confederate Railroad Successes

In spite of the numerous problems that plagued the Southern railroads, the Confederacy did at times make effective use of the system it had. In spring 1862 Union General George McClellan's Army of the Potomac landed on the Virginia coast southeast of Richmond and advanced slowly toward the Confederate capital in an offensive known as the Peninsula Campaign. By June General Robert E. Lee had command of the Army of Northern Virginia and, confident that he could fend off McClellan for a while, sent 10,000 men west by rail to strengthen Jackson in his Shenandoah Valley Campaign. This tactic so threatened the Union forces confronting Stonewall that they immediately began retreating, thereby clearing the Valley almost completely of any Federal presence. Jackson then boarded most of his army on east-bound trains to assist Lee's successful bid to drive McClellan away from Richmond in the Seven Days' Battles.

These complicated railroad maneuvers by Lee and Jackson occurred in less than a month's time, involved more than half a dozen rail lines, and resulted in troops being hauled over a total distance of 250 miles. Thanks at least in part to the steel roads, the greatly outnumbered Confederates received substantial reinforcements to help beat back two major thrusts by the Union into Virginia. These early 1862 setbacks to the North in the Eastern Theater dramatically shifted momentum to the Confederacy, guaranteeing that the contest would go on much longer and be more costly in lives and treasure.

In the Western Theater in summer 1862 Union armies captured Corinth, Mississippi, after which Federal command dispatched the Army of the Cumberland under General Don Carlos Buell to take Chattanooga, a critical rail center. Opposing Buell was the Army of Tennessee led by General Braxton Bragg who had 25,000 soldiers then located in northeastern Mississippi. Bragg countered the Union threat by sending most of his men to eastern Tennessee on a 776-mile-long circuitous route using several rail lines in just over a week. This massive operation proved to be the single largest conveyance of Confederate troops using trains in the war.

Getting Rebel infantry and their equipment to battle fronts aboard trains continued throughout the conflict. In September 1863 the "...longest and most famous Confederate troop movement by rail..." [17] took place with the transfer of General James Longstreet's army corps from Virginia to Georgia and the Battle of Chickamauga. This transport of a sizeable body of fighting men had truly spectacular results—Southern victory in the biggest and bloodiest battle ever fought in the Western Theater. The journey of Longstreet's men exemplifies two things: First, how carrying troops by train could bring triumphs on the battlefield and second, how the mounting problems in the failing railroad system kept such favorable outcomes from occurring more often.

In the first four days of July 1863 the South suffered two devastating defeats at Gettysburg and Vicksburg. Longstreet, commander of First Corps in Lee's Army of Northern Virginia, had been agitating for a fresh offensive in the West to relieve pressure in the Virginia sector. His asked that his Corps be sent to reinforce Bragg's Army of Tennessee for an attack on the Union Army of the Cumberland under General William Rosecrans near Chattanooga. President Davis approved the general's idea and on September 9 the first of Longstreet's soldiers climbed aboard trains in Virginia for the trip to northwestern Georgia.

The route taken was long—over 900 miles—and difficult. Given that the Union held Knoxville, a shorter way through southwestern Virginia and eastern Tennessee could not be used. Consequently, the trains had to roll south through the Carolinas to Atlanta, then turn northwest to Chickamauga. The journey involved a dozen separate railroads, and the numerous mismatched gauges and incomplete connections led to constant delays. At Augusta, for example, the troops had to leave the passenger cars, walk a half mile through town, then re-embark on a different carrier. In some cases, soldiers had to wait to be boated across rivers to move on.

The inferior quality of the Southern rail system meant not only frequent breakages that caused detainments but also that speed of the locomotives had to be reduced. The average speed of trains had plummeted from 25 miles per hour at the beginning of the war to an agonizingly slow six miles per hour at the end. During Longstreet's First Corps transfer, South Carolina was an especially egregious case of hindered travel; one of the artillery units still lingered in the state as the Chickamauga fighting erupted.

By 1863, the disintegration of the Confederate railroads had become acute. The dilapidated rolling stock in particular kept breaking down, with at least a quarter or more of the locomotives needing repair. The passenger cars were deteriorating rapidly, badly overtaxed by the need to transport more and more troops; military personnel made up at least one third of all riders in the first two years of the conflict. Simple plank benches hammered into the boxcars helped provide space for the flood of servicemen, who themselves roughly treated the cars they rode in. Sometimes in cold weather, they "...built fires on the floors of the freight cars and left them burning when they debarked. When Gen. Longstreet's men rode the rails to Chickamauga in September 1863, the soldiers pried boards from the side walls of the boxcars to improve the ventilation and the view" [221].

Eventually, with all the difficulties and delays, only about half of the Virginia combatants got into the fight at Chickamauga, but they decided the issue. Their well-timed charge on September 20, the second day of battle, shattered the Union lines, sending many of the Northerners fleeing from the field in disorder. Bragg, however, failed to take advantage of the rout with a vigorous pursuit of the defeated foe. Instead, he allowed the Federal army to withdraw in safety back to Chattanooga, whereupon an alarmed North opted to hurry significant reinforcements to this Tennessee battlefront. In 11 days, over 20,000 soldiers with their equipment, horses, and artillery arrived from the vicinity of Washington, a trip of over 1,200 miles involving dozens of trains. This "…was an extraordinary feat of logistics – the longest and fastest movement of such a large body of troops before the twentieth century…" [146] and a testament to the superiority of the Union's ability to prosecute war in the new Age of Railroads.

Virginia's Rail System

In 1861 Virginia possessed the most extensive railroad network in the Confederacy—about 20 per cent of the total (Fig. 13.3). Richmond and nearby Petersburg, the central area of the South's iron and steel manufacturing, also stood as the nexus of the state's rail system. Upon those tracks depended the very sustenance and survival of the capital and its sister city. The sprawling web of railroads from the south and west brought a constant flow of raw materials to feed the heavy industries,

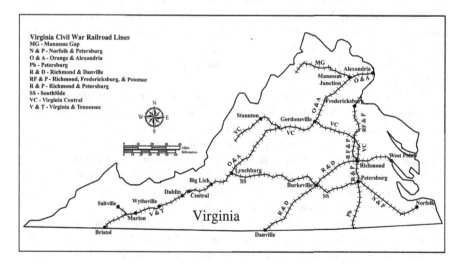

Fig. 13.3 Map of Virginia railroads in the Civil War. Note the Virginia and Tennessee ("V&T") running from Lynchburg to Bristol

foodstuffs to nourish the citizens and military personnel, and troops to man the defenses. The capture of Richmond became the principal Union objective of the struggle, and destruction of the railroad lifelines was key to the success of that strategy. When Grant at last cut the South Side Railroad on April 2, 1865, following the Battle of Five Forks just south of Petersburg, he snatched away Lee's only remaining rail connection to the south. With the South Side's loss, Richmond had to be given up and the last hopes for a free and independent Confederate States of America flickered out.

Most of the battles in the Eastern Theater of the Civil War occurred in Virginia where the dense system of railroads played a central role in the campaigns. Fourteen major companies operated in April 1861, mostly in the eastern and central parts of the state. Three of those rail lines pushed on west through the Blue Ridge, tapping into the agricultural and mineral storehouses of the mountains and valleys in the far Appalachians. The Manassas Gap Railroad, the most northern of the trio, ran from the western side of the Shenandoah Valley east through Front Royal and on to Manassas Junction. It was this carrier that rushed troops under Generals Joseph Johnston and Thomas Jackson to battle at First Manassas where they helped drive the Union army from the field. That event made the Manassas Gap the first railroad ever to bring large numbers of soldiers from a distance to an active battlefront.

At only 77 miles, the Manassas Gap was not long but had special importance in that it shipped an abundance of combatants, ordnance, and foodstuffs from the Shenandoah Valley. So much beef traveled over this road that it came to be known as the "Meat Line of the Confederacy." The junction at Manassas connected the Manassas Gap to several other lines, most notably the Orange and Alexandria. That route conducted men and materiel south to Gordonsville, an extremely busy transfer station in the central Piedmont. With so many of the big battles and campaigns of the Civil War fought near it, Gordonsville became arguably the most important rail junction in the war; some claimed that sooner or later every man in the Army of Northern Virginia passed through. From there, freight and riders could take an eastward swing directly to Richmond or proceed farther south to Lynchburg. Considering all the military resources moved by them, it is no wonder that the Manassas Gap and Orange and Alexandria ended up as two of the most fought over railroads in the Civil War.

The Manassas Gap's position in the lower Shenandoah put it dangerously close to Union-controlled territory and by spring 1862 the western half of the tracks had fallen into Federal hands. That May, as part of Stonewall Jackson's Valley Campaign, Colonel Turner Ashby and his Virginia cavalry assaulted the former Confederate line about 60 miles west of Washington. At the same time another band of gray-clad horse soldiers tore up rails and destroyed bridges farther east near Manassas Junction. Despite such ongoing Southern raids, the railway remained under Northern control for most of the remainder of the hostilities.

A second vital wartime railroad stretched across the Piedmont through the Blue Ridge and into the distant western mountains—the Virginia Central. By 1861 this 200-mile-long route had been constructed all the way from Richmond deep into the hill country west of the upper Shenandoah Valley. Jackson and Lee used Virginia

Central trains to shuttle their troops back and forth in the 1862 Valley and Peninsula Campaigns. The line intersected the Orange and Alexandria at Gordonsville, giving soldiers and supplies all-rail access to Manassas Junction and the battlefields of northern Virginia. Besides the military traffic, the Virginia Central conveyed farm products, timber, and iron from the Valley east to Richmond and other population and industrial centers. Unlike the Manassas Gap, the Virginia Central lay deep enough in friendly territory so that it steadfastly served the Confederacy for the duration of the conflict.

Nevertheless, Union raiders did strike at the Virginia Central from time to time, burning bridges and ripping up tracks. The biggest cavalry fight of the war, the Battle of Trevilian Station, broke out in early June 1864 over control of these rails. Grant, at the time pressing Lee hard at Cold Harbor north of Richmond, dispatched about 6,000 mounted troopers under General Philip Sheridan to sunder the line. Lee responded by sending General Wade Hampton's 5,000 horsemen galloping north to stop Sheridan. Hampton and Sheridan collided at Trevilian Station on June 11 and fought again next day. The Union commander chose to retire, leaving behind minimal damage that workmen soon repaired.

The Virginia and Tennessee Railroad

Of all the railroads that linked the strategic resources in western Virginia with the desperate fighting in the east, the Virginia and Tennessee ranked as perhaps the most important. President Lincoln himself called it "…the gut of the Confederacy" [168]. It functioned as the only line that could get the immense stores of mineral riches in southwestern Virginia to the furnaces and foundries in places like Lynchburg, Richmond, and Petersburg to be transformed into weapons of battle. But far more than mineral resources moved over these rails. Much like the Shenandoah Valley, Southwest Virginia served as a prime Confederate breadbasket. As such, the Virginia and Tennessee transported huge amounts of wheat, corn, potatoes, and other vegetables to feed the soldiers and civilians mostly in the east, although some of the produce went west as well. Indeed, most of the bread for the Confederacy's eastern armies came from southwestern Virginia. Livestock, in particular beef cattle and hogs, and the salt to preserve the meat once it was butchered traveled in great quantities along the Virginia and Tennessee tracks.

As the fighting dragged on, it became clear that Confederate final victory depended on having a lasting ability to keep the fighting forces provisioned and armed. Eastern Virginia held the South's greatest concentration of manufacturing and its main army, and southwestern Virginia had minerals and food that could supply both. The Virginia and Tennessee bridged the gap between the two regions. Losing the railroad would be a disastrous blow to the Confederacy, and the North soon came to realize this.

Desire to build a railroad through the southwestern quarter of the state appeared in the decades before the war. Although remote and thinly populated, by the early nineteenth century the region had established a modest economy based on its minerals and agriculture. This attracted the attention of powerful Tidewater politicians and businessmen who sought to tap into the underdeveloped assets of Southwest Virginia. During the 1820s and 1830s interest increased in connecting eastern and western Virginia by the building of new roads, canals, and ultimately railroads. In March 1849, the state General Assembly voted to fund construction of the Virginia and Tennessee Railroad to extend from Lynchburg to Bristol. Work got underway in Lynchburg the following January and ended in Bristol in October 1856.

The railroad through southwestern Virginia emerged as one of the lengthiest in the entire South. From Lynchburg it traveled west across the gently rolling hills and valleys of the Piedmont, then through a gap in the Blue Ridge at a hamlet called Big Lick (today's Roanoke). There the tracks turned southwest and followed the Great Valley to Bristol, where they connected with another line that maintained the route to Knoxville and Chattanooga. The Virginia and Tennessee extended for 213 miles, making it the longest railroad in the state and the longest line of uniform gauge in the Confederacy. Structures along its length included five tunnels, 233 bridges, and 19 depots. The bridges were all short and over small streams and creeks except for one—the 730-foot-long bridge over the New River at Central Depot.

The valley railway was also one of the most carefully constructed and expensive in the nation. Many railroads of that era ran on tracks laid directly over the natural ground surface whereas the Virginia and Tennessee had a substantial bed of stone ballast supporting every rail joint. An extensive system of drainage ditches resulted in fewer washouts of the roadbed than happened on other lines. When finished, the route cost $84,000 per mile to fabricate, a cost significantly higher than most other places in the South or North.

Marked changes to southwestern Virginia came quickly after the iron rails forged links to markets in Tidewater as well as the entire southeastern United States and beyond. Agricultural yields surged. Growers, for example, heretofore raised tobacco as a minor crop but production soared over 2,000 per cent from 1850 to 1860, due in no small part to the Virginia and Tennessee. The scale of farming endeavors switched from mostly subsistence yeoman plots to greater numbers of large commercial ventures managed by wealthy landowners. The mineral industries boomed as well. Lead and shot from the Wythe County mines, salt and gypsum (plaster) from the Saltville area in Smyth County, and iron from furnaces scattered throughout the region all saw expanded output. Additional copper mines opened and renewed interest in coal mining appeared, a harbinger of the ascent to dominance of this industry in the postbellum late 1800s.

The steel tracks fostered social and political changes even more consequential than economic ones. The rise of expansive, more plantation-like farms in the open spaces of the Great Valley demanded many more black laborers. From 1850 to 1860, the number of slaves in southwestern Virginia went up by 2,584, a nearly 16 per cent growth rate. The Virginia and Tennessee itself was built mainly by slave labor; in 1856, hired slaves accounted for 435 of the 643 workers on the line. The spread of slavery in Southwest Virginia helped enormously to bind this section more strongly to the culture and commerce of central and eastern Virginia, where talk of secession grew louder as the 1860s dawned.

By early 1861, the election of Abraham Lincoln to the presidency that previous fall had prompted five Southern states, fearful for the survival of slavery and the way of life it supported, to withdraw from the Union. Virginia, including the counties of the southwestern part of the state, followed them in rebellion five days after civil war burst forth at Fort Sumter on April 12. Union sentiment remained strongest in the mountainous northwestern counties where no Great Valley existed to promote development of broad agricultural estates with numerous slaves. Two years later, the mountain people broke away to create the free state of West Virginia.

In the end, the far southwestern counties stayed in Virginia and the newly formed Confederate States of America, the ties holding them there in no small part fashioned by the steel sinews of the Virginia and Tennessee. Now, with the onset of war, that railroad would become a prime target of the Union forces presently gathering to invade Virginia.

Notes

The 1854 Virginia and Tennessee celebration at Central Depot is from Worsham [278]; the brief remarks about the Civil War attack to cut that line at Central are distilled from McManus [144]. In the "Iron Horse" section of the chapter, the early development of railroads, particularly in Great Britain, is from Wikipedia [263, 266] sites on the history of rail transport and the timeline of railroad history. The discussion of the explosive growth of railroads in the antebellum United States is from McPherson [147] and the impact of the iron rails on Civil War military actions is drawn from Black [18] and Hawley [78]. Concerning the "Deficiencies of the Rebel Railroads" section, I found the information on the great disparity between Northern and Southern railways in Ketchum and Catton [112] and McPherson [147]. The effect of slaves and cotton as Southern disincentives to railroads and other forms of industrialization is from Atack and Passel [11] and comments from my reviewer Dr. Bill Grant. The rest of this section and the following one on "Confederate Railroad Successes" are based largely on Black [18], still the single best source on the Southern rail lines, and Stover [222], a very good short history of the Confederate railroads. The most important reference for the story of "Virginia's Rail System" is Johnston [100]. In addition to Johnston's book, I used material from online write-ups for the Manassas Gap Railroad in Gray [70] and the Virginia Central in Bocian and Salmon [19]. The examination of "The Virginia and Tennessee Railroad" is derived first and foremost from Noe [169], the essential source of information about this rail line. All of the comments on the political and social impact of the Virginia and Tennessee are from him. Johnston [100] and McLean [140] provided useful information here also.

Chapter 14
Union Raiders in the New River Valley

"To cut New River Bridge and the railroad...would be the most
important work"

The issue was settled in fighting that lasted just under one hour: an invading United States force would advance deeper into southwestern Virginia to attack the Virginia and Tennessee Railroad. The Union victory making this possible came about on Monday, May 9, 1864, at the largest battle ever fought in the southwestern quarter of the state—Cloyds Mountain in Pulaski County (Fig. 14.1). One Northern veteran of the struggle noted that "Most of the regiments engaged in it had served in greater and more important battles, but all united in the opinion that, for fierce and deadly intensity, Cloyd[sic] Mountain exceeded them all" [269]. On this beautiful sun-splashed day in the Appalachian Mountains, roughly 9,000 soldiers clashed and 1,226 became casualties. Union killed, wounded, and missing amounted to about ten per cent of their strength and Confederate losses approached an appalling 23 per cent.

Next morning, again under gorgeous spring skies, Northern and Southern troops collided when a ferocious cannon duel broke out ten miles away at the New River Bridge near Central Depot. For nearly three hours artillery blazed away from opposite banks of the ancient stream to decide whether the vital railroad span would survive or be destroyed. Fighting for the Federals at Cloyds Mountain and New River Bridge were some noteworthy people. Colonel Rutherford B. Hayes and Lieutenant William McKinley would go on to hold the nation's highest political office. A trooper in Hayes' brigade killed at the New River Bridge—we do not know her name—remains one of the very few females documented to have fallen in combat during the Civil War.

© Springer International Publishing Switzerland 2015
R.C. Whisonant, *Arming the Confederacy*, DOI 10.1007/978-3-319-14508-2_14

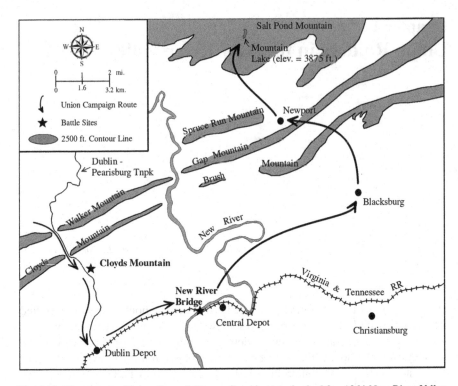

Fig. 14.1 Map showing Union General George Crook's route in the May 1864 New River Valley Campaign. Crook's Kanawha Division fought the Confederates at Cloyds Mountain on May 9 and the New River Bridge on May 10, then retreated back to West Virginia via Blacksburg and Mountain Lake

The Federal Advance into Southwest Virginia

General Ulysses S. Grant acceded to command of all the armies of the United States in March 1864. Grant intended to crush Lee in Virginia by force of arms in major battles and by using multiple encircling Federal units to cut off supplies and rein-forcements from other areas. To execute his grand strategy, the new Commander-in-Chief soon had the Union on the march throughout the state. General Ben Butler moved on Richmond up the James River from Fortress Monroe in the Hampton Roads region. General Franz Sigel advanced south in the Shenandoah Valley aim-ing to reach Staunton, then thrust beyond to Lynchburg. Grant himself joined the Army of the Potomac, commanded by General George Meade, in the field and pushed down from Washington toward the Confederate capital. The fourth compo-nent of the closing trap involved a strike from the southwest with General George C. Crook's Kanawha Division stationed in Charleston, West Virginia (Fig. 14.2). Crook received orders to proceed through the Allegheny Mountains and sever the Virginia and Tennessee Railroad in the New River Valley. If possible, the massive salt installations at Saltville were to be decimated as well.

The idea of an attack on the Virginia and Tennessee in the New River Valley did not originate with Grant. As early as 1861, Union military staff concerned with

Fig. 14.2 Union General George Crook. Photograph courtesy of the Library of Congress. In May 1864 Crook won the Battle of Cloyds Mountain and burned the New River Bridge, but the effects of his campaign were short-lived

southwestern Virginia, in particular Colonel Hayes, planned operations to cut the railroad by demolishing the New River Bridge. In 1862 and 1863, a number of ill-conceived and abortive attempts to do just that took place. The deepest of those incursions happened when Hayes got his command within 15 miles of the railroad in May 1862 before being driven back into West Virginia.

Unlike these earlier fiascos, Crook's 1864 expedition had real strength behind it and still bigger strategic goals. If Crook succeeded in his assault on the rail line, he would press on to Lynchburg, linking up with Sigel's army advancing from the Shenandoah to attack the city. Lee would therefore be completely isolated from desperately needed resources trying to get through from the west. At the same time, cutting the Virginia and Tennessee would prevent Lee and his men from using its trains to escape into the harsh landscape of southwestern Virginia where they might pursue guerilla warfare.

Given the lack of sizeable commands provided for operations in the mountainous terrain of this sector, fast-moving smaller detachments offered the most effective option for Federal officers to get at the railroad and mineral works. Destroying an entire railroad is not feasible for such units, thus depots and bridges became the principal targets. Without question, destruction of the Long Bridge over the New River at Central Depot offered the surest way to disable the railroad. Union strategists also had great interest in nearby Dublin Depot, headquarters of the Confederate Department of Southwest Virginia. Significant military personnel and provisions, in addition to considerable railroad facilities, resided there. Even so, Grant recognized that above all, the span over the river at Central had to come down, writing to Sigel on May 2 "To cut New River bridge and the railroad ten or twenty miles east from there would be the most important work Crook could do" [234].

On April 29, Crook launched his invasion, leaving Charleston with 6,155 infantry, supporting artillery, and 2,500 cavalry. Colonel Hayes, an Ohioan who had served in western Virginia from the early days of the conflict (and, since June 1863, in the newly formed state of West Virginia), led one of Crook's three infantry brigades (Fig. 14.3). Within his outfit, Hayes had a young lieutenant in the 23rd Ohio named William McKinley (Fig. 14.4). The two future presidents campaigned together

Fig. 14.3 Union Colonel
Rutherford B. Hayes.
Photograph from the Civil
War Photographs Collections,
U. S. Army Military History
Institute. Hayes served with
distinction in the Civil War
and afterwards was elected
president in 1876

Fig. 14.4 Union Lieutenant
William McKinley.
Photograph from the West
Virginia State Archives, Boyd
B. Stutler Collection.
McKinley, the last Union
veteran to be elected
president, was assassinated at
Buffalo, New York, in 1901

throughout the New River expedition. Ironically, McKinley had fought previously at Antietam and survived the carnage at Cloyds Mountain, only to fall to an assassin's bullet in Buffalo, New York, near the beginning of his second term.

Crook's men moved jauntily up the Kanawha Valley that first day, cheered by Hayes' and several other officers' wives who trailed behind them briefly in a chartered stern wheeler on the river. The Federals left the Kanawha Valley on May 2 and, pelted by sleet and snow, began the arduous trek over the rugged Allegheny Mountains. On the same day, Crook instructed cavalry commander General William Averell to assault the Saltville operations, then proceed to Dublin, wrecking the railroad along the way.

On May 6, as Lee and Grant continued a second round of bloody battle in the Wilderness far to the east, Crook's column met and rapidly overwhelmed a small group of Confederates near the Virginia border. The next day the Kanawha men entered the Valley and Ridge of Virginia, driving Rebel skirmishers before them. On the evening of May 8, after long hours of marching under increasing fire from bushwhackers, Crook's division made camp under clear starry skies just northwest of a steep ridge known as Cloyds Mountain. Only two miles away, across the mountain top to the southeast, Southern General Albert Jenkins commanded a cohort of infantry and artillery hastily gathered to blunt the Northern threat (Fig. 14.5). Jenkins' troops sat entrenched atop low hills at the base of Cloyds Mountain behind barriers of logs, fence rails, and earth. In front of these works, between the Confederates and their Union adversaries, a little stream named Back Creek flowed through an open grassy valley. As night gathered, the gray-clad defenders waited anxiously, determined to fight to the death for Dublin and the railroad five miles to their rear. Tomorrow, many on both sides would be called upon to do that very thing.

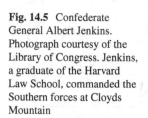

Fig. 14.5 Confederate General Albert Jenkins. Photograph courtesy of the Library of Congress. Jenkins, a graduate of the Harvard Law School, commanded the Southern forces at Cloyds Mountain

The Battle of Cloyds Mountain

Crook's soldiers moved out of camp at dawn on May 9. Expecting to meet the main body of Confederates along the mountain top, the blue coats advanced cautiously through the tangled brush and woods up the northwestern slope of Cloyds Mountain. But only a few pickets confronted them at the ridge crest and a crisp exchange of fire swiftly drove the Rebels off. Crook, upon reaching the mountain's summit, observed the half-mile long Confederate defenses before him at the foot of the mountain just beyond Back Creek and opined: "The enemy is in force and in a strong position. He may whip us but I guess not" [141]. At this point, the Union general deployed his troops into a line parallel to the ridge top and directed them to descend the slope toward their opponents (Fig. 14.6).

The Federals by now numbered 6,555 men organized into three brigades with 12 pieces of artillery. General Jenkins had around 2,400 regulars with ten cannon to meet the approaching assault. The battle proper commenced about 11 a.m. when a brisk round of cannon fire broke out, succeeded by the crackle of musketry. Crook initiated his attack with his Second Brigade advancing against the Confederate right. At the same time, he ordered that as soon as heavy firing on the Union left announced the onset of a serious engagement, the other two brigades should strike the Rebel center and left.

As the Federal move forward got underway, a terrific roar of cannon and rifles erupted all along the battle front. Confederate musket fusillades and clusters of grape and canister from the artillery swept the field, tearing huge holes in the blue columns.

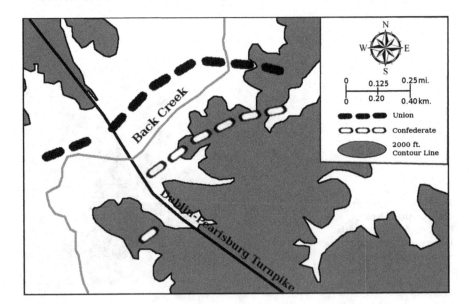

Fig. 14.6 Map showing the positions of the Union and Confederate lines at the Battle of Cloyds Mountain, May 9, 1864. The positions are generalized and represent the situation at the start of the battle

Crook's infantry in the Second Brigade suffered terribly and some units were pushed back, leaving the ground "…thickly strewn with their killed and wounded" [233]. Dry leaves and grass covering this part of the field caught fire and cremated several of the Union casualties before they could be rescued. The brief success of the Confederate right wing against the Second Brigade caused the Southerners to cheer triumphantly as they considered the victory won, but that was not to be the case.

The Union First and Third Brigades now smashed into the Rebel center and left. Hayes commanded the First Brigade and urged his men on into the fray. Hayes, a lawyer in Ohio before the war, had already proved to be a good officer, effective in the field and much admired by his men. One of his soldiers described him as a leader who, once the fighting began, became inflamed with excitement such that his face fairly glowed with the light of battle. The colonel performed well this day, bringing his men out of the woods on the lower slope of the mountain and leading them across the bottomland meadow between them and their enemy. Volley after volley ripped into the Ohioans, yet on they rushed, crossing Back Creek and reaching the base of the hill where the main Confederate lines perched.

General Crook himself, caught up in the thrill of the charge, joined the First Brigade racing across the meadow and splashing through Back Creek, later said to have run red with blood. Crook, like many general officers in those times, often wore large riding boots. As he waded across the creek, the boots filled with water; the men close by had to drag him from the creek while bullets flew all about them.

Once across the stream and floodplain, officers reformed the Northern soldiers, who pushed their way up the slope of the ridge and closed with the barricaded Confederates in furious combat. This brought the struggle to its murderous peak as frenzied warriors fought hand-to-hand with clubbed muskets, bayonets, knives, and fists. Wounded and dead on both sides piled up along the logs and rails breastworks. Still the Union pressed the attack, and the gray wall began to buckle. Watching from nearby astride his horse, General Jenkins tried to rally his men, then he too went down when a musket ball shattered his upper left arm. He would die a few days later in a Federal field hospital when a careless attendant knocked loose the ligature from the amputated limb and he bled to death.

With Jenkins fallen and his infantry beginning to leave the field in confusion, Colonel John McCausland assumed command (Fig. 14.7). Suddenly, the Confederate right wing collapsed, followed by the entire position cracking apart. Men and artillery hurriedly gave up the prepared defenses and retreated across the open ground behind them. Some troops started to flee toward Dublin in panic, but McCausland rallied his charges and fought a skillful rear guard action. Finding a low hill top, the Confederates wheeled about and took on their pursuers briefly in an engagement called the Battle of the Second Crest. Shortly, with much of his artillery and walking wounded safely off the field, the Southern commander ordered retreat to Dublin.

The May 9, 1864, Battle of Cloyds Mountain did not last long—according to Virginia gunnery sergeant Milton Humphreys, "…the time that elapsed from the first heavy roar of musketry until the artillery and thin line of infantry were withdrawn was just fifty-two minutes" [92]. The combat was, however, exceedingly savage with quarter frequently not given by either side. At times amid the fighting, some enraged soldiers

Fig. 14.7 Confederate
Colonel John McCausland.
Photograph courtesy of the
Library of Congress.
McCausland assumed
command at Cloyds
Mountain when General
Jenkins fell wounded

bayoneted unarmed foes that had already surrendered. A local Presbyterian minister
dressed in civilian finery complete with high silk hat fought and fell wounded. When
Union medics came upon him propped up against a tree, they gave him no medical aid
and left him to eventually die because he was not in uniform.

The high casualty rates attest to the intensity of the struggle. The Northerners lost
688 men and Southern casualties totaled 538. Among the Rebel dead lay Captain
Christopher Cleburne, younger brother of famed Western Theater Confederate
General Patrick Cleburne. The 21-year-old Captain Cleburne was hit while mounting
a countercharge against the blue waves sweeping down the Dublin-Pearisburg
Turnpike. This bold tactic slowed the onrushing enemy and helped save McCausland's
retreating units from real disaster. Cleburne went down fatally wounded on the pike
and asked to be buried by the roadside, where comrades granted his wishes. Today his
lonely gravestone keeps vigil over the battlefield at a highway wayside named in his
honor. Within a few months after the Cloyds Mountain battle, Patrick Cleburne would
die in the Confederate debacle at Franklin, Tennessee.

Artillery Duel at the New River Bridge

Shortly after the battle, McCausland's men reached Dublin, gathered what supplies
they could, and then abandoned the rail town later that afternoon. Moving east with
his remaining command, the Confederate colonel determined to make a stand at

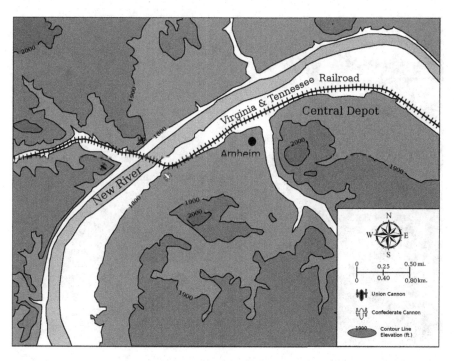

Fig. 14.8 Map showing the location of the New River railroad bridge where the May 10, 1864, artillery duel was fought. Cannon positions are approximate. Arnheim, the home of Dr. John Radford, was shelled by the Union

New River Bridge about ten miles distant at Central Depot (Fig. 14.8). McCausland sent his artillery and supply wagons to Central along the macadamized Southwest Turnpike, the main Valley highway. The cannon and wagons crossed the New River at the Ingles Ferry Bridge just upstream from the railroad span, and proceeded down the river to take up positions at the south end of the rail bridge. Next morning, McCausland had the Ingles Ferry bridge burned to prevent Union use. The Southern infantry retreated from Dublin along a separate route from the field pieces and wagons, traveling instead along the railroad to the approaches of the New River Bridge on the north bank of the river. There, high on the tall dolomite bluffs, a few cannon had been pre-positioned in defensive earthworks overlooking the rail span.

But McCausland now decided not to use those fortifications. Not wanting to have his command trapped with a river to its back in case of defeat, he concentrated his troops and big guns on the south side of the New. Around dusk, his infantry got across the river walking along the railroad bridge. Once over the stream they reformed, while some sharpshooters and skirmishers spread out along the river bank near the span. Later that evening, the Virginians brought most of the cannon from the north side of the river across using a single flatboat; however, the two heaviest pieces could not be moved and had to be spiked and left behind. For several hours, squads of soldiers trudged across the railroad bridge carrying ammunition for the batteries in nosebags and haversacks. It was after midnight before McCausland at last had his reunited forces arrayed along the brow of a gentle hill near the south end

Fig. 14.9 View looking north across the modern New River Railroad Bridge. Federal cannon were sited on the two ridges east and west of the railroad in the middle distance

of the bridge. Here the Rebels watched for Crook's men whom they knew would surely attempt to destroy the only means the Virginia and Tennessee trains had of crossing the New River.

Meanwhile, not long after the Confederates had withdrawn from Dublin following the fighting at Cloyds Mountain, the Kanawha Division occupied the town and bivouacked for the night. Next morning, May 10, dawned sunny with warm breezes, and the Federals spent several hours putting military and railroad supplies in Dublin to the torch before moving out to Central Depot. Crook's men advanced along the railroad, destroying tracks as they progressed. Around 10 a.m., the forward skirmishers reached the high ground near the north end of the bridge (Fig. 14.9). Confederate batteries immediately opened fire as the Union came up quickly in force. Fourteen Southern cannon opposed 12 Northern guns, but the Union weapons were of better quality. Over most of the ensuing three hours, artillery roared to determine the fate of the New River Bridge.

Central Depot was a modest yet growing railroad community in 1864 with about 20 families living there. The men worked mostly for the Virginia and Tennessee, which had erected significant rail yard facilities, including a roundhouse that typically contained a number of cars and engines. The little town lay nestled along the south side of a large meander loop of the New River. The bedrock geology of this locality is dolomite mixed with some limestone, rock types that groundwater readily dissolves to generate an abundance of caves and caverns throughout the region. As time passes, the roofs of these solution cavities fall in, producing semicircular depressions in the ground surface called sinkholes. These small declivities are especially numerous on the north side of the river where dolomite exposures abound and steep rock bluffs rise about 200 feet above the water level. Here along the high cliffs on ground dotted with sinkholes, the Union sited its artillery, placing some pieces in the empty Confederate earthworks.

South of the river, elevations are generally less where the coursing water has eroded away the dolomite and deposited a covering of silty sediments to form a floodplain. Several older floodplains, flat stretches of ground known as terraces, mark where the river in former times flowed at higher elevations but has now cut down to its present lower level. These abandoned ancient surfaces of river erosion rise like giant stair steps south from the modern floodplain. On one of the higher terraces overlooking the New is Arnheim, the still-standing stately home of local physician Dr. John B. Radford, for whom Central Depot would later be renamed. Built in 1836, Arnheim received several hits from cannon balls when Colonel Hayes directed fire on it, mistakenly assuming this impressive structure to be enemy headquarters.

The geology of the Central Depot-New River Bridge vicinity had several note-worthy effects on the cannon duel that occurred. First, because Federal gunners held the higher ground created by the rocky bluffs, they were able to watch the entire path of their shells until detonation. Southern cannoneers could see neither the flight nor landing of their missiles once they disappeared over the top of the cliffs, a decided disadvantage in an artillery exchange. Second, one of the few casualties incurred by either side during the barrage took place when a trooper under Hayes refused his order to take refuge in one of the sinkholes near the north end of the bridge. The colonel recorded the event in his diary:

> There was a large lime stone sink hole in which I ordered the men to lie down. All obeyed promptly except one dismounted cavalryman who in a pert and saucy way turned to me and said, 'Why don't you get off your horse too?' On my repeating the order, the cavalryman replied, 'I'll get down when you do.' Just as I was insisting on his obeying the order a shell burst near us—the cavalryman was fatally and shockingly wounded and was then discovered to be a woman. She died almost instantly. [96]

Another geologic feature came into play after the battle when the Confederates had withdrawn. A few Union soldiers crossed the river and entered the Radford home looking for plunder. No looting or vandalism took place, however, thanks to the family guns and silver being hidden the day before in a limestone cave within a wooded hillside near Arnheim.

When Crook's troops and artillery arrived in strength on the north side of the river on May 10, they deployed into positions atop the steep cliffs on both sides of the bridge. As the big guns began to thunder, the concussion from a cannon shell bursting very near the Union general "...had such a sickening effect upon him that he was forced to dismount" [49]. Seeing the riderless horse dashing frantically for the rear, "...the rebels raised an exultant yell" [49]. The joyous gunners thought they had killed Crook, but the shot stunned him only temporarily.

While this was going on, a Northern officer had noticed railroad workmen busy at the Central Depot roundhouse rushing about to remove engines and cars from harm's way. A rifled cannon at once redirected its fire at this new target. Although the roundhouse stood at least a mile and a half distant, one round scored a direct hit that exploded within the building "...seriously interrupting their operations and scattering the Johnnies [Rebels] in every direction" [8].

As the battle raged on, a Southern battery lofted a 20-pound shell with a burning fuse into a Union gun emplacement where it rolled under an artillery limber. One of the crew, Private John Wilhoff "...was preparing ammunition for a gun, the burning fuse facing and sputtering right near his ammunition. Without hesitation Wilhoff caught the shell, cut out the burning fuse and tossed the shell from him, perfectly harmless" [8]. This brave action saved "...the lives of many; for had the shell exploded, it would have blown up the limber and done great damage..." [8] to other soldiers posted nearby.

Behind the Northern artillery an entire brigade of infantry had been positioned in a dense stand of woods. These men endured the enemy barrage "...holding down roots and scrambling out of the way of falling limbs and treetops, caused by the exploding shells" [8]. Two companies of West Virginians were posted as skirmishers in front of the Union cannon. Shortly, these soldiers and some Pennsylvania men advanced toward the river as the Confederates ran short of ammunition and Crook ordered the bridge to be burned.

From the south side of the river, the Rebels had kept up a continuous bombardment as enemy shells landed among them. One of the battery commanders, Captain Thomas Bryan, encouraged his gunners by riding his white horse along the row of cannon. This daring act quickly drew the attention of Crook's artillerists; one of their shots killed Bryan's horse and threw him from the saddle. His troops carried him from the field. Attracted by the bellow of battle, a group of local young people gathered to watch the encounter. Union observers saw this and a well-placed round of fire suddenly panicked the ladies and gentlemen and sent them flying. Writing from the Southern perspective, Sergeant Milton Humphreys recorded that "During the cannonade the Federals made repeated attempts to burn the bridge, but were always thwarted by some of the artillery and the skirmishers and sharpshooters" [93].

Shortly after noon, Confederate fire slackened as ammunition stores ran low. Union troops now moved speedily to carry out their general's command to set the bridge ablaze. Captain Michael Egan procured matches, climbed onto the structure, and soon flames surged through the dry wooden timbers. The regimental band of the 23rd Ohio played a rousing martial air as the victors, waving flags and shouting jubilantly, lined the tops of the stony cliffs to watch the collapsing span plunge into the New River.

The sturdy bridge piers, however, had survived the shelling relatively unharmed. When asked about materials for blowing up the bridge foundations, Crook replied that it was the intention to have brought explosives, yet somehow they had been overlooked. The general then directed that solid shot be used on the pilings, but these had no effect on the solid structures. About 1 p.m., as burning pieces of timbers floated away downstream, McCausland broke off the fight completely. His ammunition nearly exhausted, the Confederate commander withdrew east along the rail line to Christiansburg, planning to defend the Virginia and Tennessee there.

Retreat Back to West Virginia

Despite his opponents quitting the fight, Crook had other ideas besides campaigning any farther into unfriendly territory. His prime goals of wrecking Dublin and burning the New River Bridge had been accomplished. The Union leader was especially worried by his troops' discovery of Rebel telegraph dispatches the day before in Dublin indicating that Lee had repulsed Grant in the Wilderness. But this was erroneous information. Although Lee had indeed inflicted terrible losses on the Army of the Potomac, its tough minded new commander had no intention of retreating. Rather, Grant began a side-stepping advance toward Richmond—the Overland Campaign—that led to the final denouement with the Army of Northern Virginia. Crook knew none of this, instead believing that Confederate forces of unknown strength and location were now free to trap him deep in hostile country. Consequently, he chose to fall back to his base at Meadow Bluff, West Virginia, as rapidly as possible.

In the meantime, cavalryman Averell had run into troubles of his own. Dispatched by Crook to attack Saltville, Averell opted not to challenge the strong defenses there under John Hunt Morgan. But Morgan came after Averell, and, on the same day that cannon fire boomed at Central Depot, caught him a few miles north of Wytheville. A running battle ensued that ended when the Union horsemen withdrew. Next day, May 11, Averell arrived in Dublin only to find Crook already departed from the New River Bridge. When a detachment of Averell's cavalry finally joined Crook later that evening, the only plan in either general's mind was to fight his way back across the Valley and Ridge to the sanctuary of West Virginia.

Upon leaving Central Depot after incinerating the Long Bridge, Crook crossed the New River a few miles downstream at Pepper's Ferry and stopped for the night. Before long, crowds of singing and dancing slaves found them and stayed, intent on walking with the troops out of bondage. Morning brought more exuberant blacks joining the march to the North and freedom as the Kanawha Division pushed on. Arriving in Blacksburg, many of the men encamped on the grounds of the Olin and Preston Institute. Crook and his officers spent the night in one of the campus buildings of the school that after the war became the Virginia Polytechnic Institute, today's Virginia Tech.

The following morning, May 12, Crook's forces moved out of Blacksburg before dawn in a driving rain. A few miles north of town, they crossed over two prominent sandstone ridges—Brush Mountain, the northeastern extension of Cloyds Mountain, and Gap Mountain. The Union raiders had now left the Great Valley and entered the much more precipitous western Valley and Ridge. Descending into a narrow vale, the Federals encountered a thin Confederate detachment at Newport that was quickly driven away. The muddy slog homeward dragged on as Rebel skirmishers and snipers harassed the Northern column and the awful weather would not let up.

Late in the day of the 12th, the Kanawha troopers struggled up Salt Pond Mountain, at 4,300 feet one of the highest peaks in western Virginia. The climb proved incredibly taxing; exhaustion soon overtook the men, wagons broke down, and mules fell over in their tracks. Colonel Hayes wrote in his diary: "A horrible day, one of the worst of my experience" [142]. That night the advance portion of Crook's weary warriors set up camp at Mountain Lake near the summit of Salt Pond Mountain (Fig. 14.10). Behind them, many of the bedraggled men simply dropped down where they were, a long blue line strung out along the dreadful mountain road.

Mountain Lake, the "Silver Gem of the Alleghenies," already enjoyed fame as a popular spa and resort when Crook's troopers saw it in 1864. This small water body is also one of the most unusual geological phenomena in all of the southern Appalachians. The mountain top pool is the only natural lake in the unglaciated Valley and Ridge terrain south of Pennsylvania, and one of only two natural lakes in Virginia. (Lake Drummond in the Great Dismal Swamp in the Coastal Plain is the other.) The lake towers nearly 1,000 feet above the surrounding ridge tops and 2,000 feet above the New River less than six miles away.

Fig. 14.10 Mountain Lake atop Salt Pond Mountain in Giles County, Virginia. The Union camped here overnight during the retreat from the Cloyds Mountain and New River Bridge actions

No one knows exactly how Mountain Lake formed. Over the years, suggestions have included: ground water dissolving away the bedrock; an Ice Age glacier scouring out a depression; volcanic explosions or meteorite impact (neither very likely); and even cows stomping around a salt lick on the mountain top. A more recent and plausible idea is that some sort of landslide dammed up a creek valley, thus creating the reservoir to hold waters entering from the stream flow and groundwater springs. The lake sits near the center of one of the largest earthquake-prone areas in the eastern United States, the Giles County Seismic Zone. (Quakes with Richter magnitudes as high as 5.8 have originated here.) Perhaps ancient seismic shocks caused the soil and rocks on the slopes around the lake basin to slide downhill and block any outlet for the valley water flow. At present, the precise origin of this truly unusual water body remains enigmatic.

Descending the northwestern slope of Salt Pond Mountain on May 13, Crook's soldiers continued their tortuous journey in drenching rain through the Allegheny high country. Skirmishing along the way, the Federals crossed a succession of sandstone peaks over 4,000 feet in elevation before coming down into the Greenbrier Valley and the relative safety of West Virginia. As the bone-tired marchers reached the Union base at Meadow Bluff on May 19, the sun finally broke through. It was the first sunshine they had seen in nine days.

Averell's soldiers beat their way to sanctuary as well. Upon leaving Dublin on May 12, the cavalrymen proceeded to Christiansburg where they ripped up several miles of railroad tracks, burned several buildings and a water tank, and destroyed a considerable amount of railway and military supplies. Turning northwest and passing through Blacksburg, Averell's command labored its way up and down the same high ridges traversed by Crook's men just a short time earlier, albeit by a partially different route. The beleaguered troops finally rejoined Crook's division on May 15 in Union, West Virginia; all were safe and reprovisioned when supply trains met them on May 20 and 21 in Meadow Bluff. At last, the grueling campaign to sever the Virginia and Tennessee Railroad had ended.

Significance of the Railroad Raid

But just how important was the May 1864 railroad raid into the New River Valley? Certainly it was physically demanding for the invading troops who had to travel over some of the most difficult terrain in the Appalachian Mountains. Historian Howard McManus observed that "For twenty-one days, Crook's army had marched 270 miles through 11 counties. Seventeen rugged mountain ridges and countless streams impeded their progress. For sixteen days, heavy storms plagued the column" [143]. Crook considered the expedition a success in that he demolished facilities and supplies at Dublin and brought down the New River Bridge. On the other hand, at least one critic pointed to his failure to achieve the greater strategic goals of linking up with other Federal armies to choke off Lee, concluding that the incursion "...ended with results hardly worth the powder" [143]. General Grant, nevertheless, seemed pleased, later writing: "Crook and Averell advanced from the Gauley at the appointed time, and with happy results. They reached the Virginia and Tennessee railroad at Dublin, destroyed a depot of supplies, tore up the track and destroyed New River Bridge" [97].

In the end, the main accomplishment of the attack—the seemingly successful demolition of the bridge at Central Depot—really wasn't of long term significance. Only the wooden superstructure had burned, leaving the foundation piers in the river completely intact. Within five weeks of the artillery duel, Confederates had the span repaired well enough so that trains rumbled over the New River once again. The workmen this time used flame-resistant green wood to rebuild the structure. During a later Union attempt to fire the bridge once more, the timbers would not ignite. The Long Bridge survived to the very end of the hostilities.

Notes

Although this chapter on the May 1864 Union campaign to cut the Virginia and Tennessee Railroad is divided into a number of sections, many of the sources cited below were used throughout the chapter. Foremost among these are two very good books on the topic—Johnson [98] and McManus [144]. These accounts have much detail on both the Cloyds Mountain and New River Bridge battles with numerous primary sources referenced. Works by Smith [216], Walker [246], Marvel [131], and Robertson [192] also provided information about the two battles. Among the firsthand accounts I examined are those from Egan [50], Arthur [9], the *Official Records* [233, 234], and Wilson [270] for the Union perspective and Humphreys [94] for the Confederate side. McLean [140] was used for information on the early (pre-1864) strikes against the railroad in Southwest Virginia, and particularly Colonel Rutherford B. Hayes's key role in the conception and planning of those attacks. Werrel [257] is a look at the entire Civil War history of the 36th Ohio Infantry regiment, part of General George Crook's command for most of its service; his Chap. 5 is devoted to the Cloyds Mountain and New River Bridge actions. Finally, the general geology of the campaign area was abstracted from, Dietrich [45], Frye [68], Schultz and Bartholomew [207] and the geology specifically for Mountain Lake from Mills [152], Whisonant and Watts [260], Cato et al. [30], and Freeman et al. [65].

Chapter 15
Epilogue

"You have done all that can be done"

When the armies at last stood down, the mineral industries and Virginia and Tennessee Railroad in western Virginia, all damaged yet none completely destroyed, faced uncertain futures. The years immediately after peace came were difficult, but eventually capital flowed in from Northern and foreign financiers, particularly British. With the fresh funding, most of the operations revived to become key elements in rebuilding the post-war economy. By the 1880s, southwestern Virginia's iron and coal production boomed as the railroad network, chief carrier of this mineral wealth, underwent significant expansion. Even though limited salt manufacture continued at Saltville and better days lay ahead, the prodigious Civil War output of 10,000 bushels a day was never reached again. Salt making ceased altogether in 1896, replaced by a prosperous salt byproducts industry. At Austinville, zinc processing ultimately outstripped lead in the late 1800s, and the mines turned out metal for another century.

Niter

The lone exception to the postbellum successes of the mineral industries proved to be the mining of niter in the caverns of western Virginia. Two factors caused its demise. First, the end of the Civil War collapsed the demand for gunpowder and the potassium nitrate used to make it. Even more important, just as the fighting concluded, Swedish chemist Alfred Nobel solved the problem of the dangerous instability of nitroglycerine by mixing it with inert clay, bringing in a new era of high explosives innovation. Nobel's dynamite carried a blast three times more powerful than saltpeter-based gunpowder, now called black powder, and rapidly assumed command of the explosives market. In the 1890s widespread adoption by the world's militaries of smokeless gunpowder made with nitrocellulose spelled the end of black powder as a

© Springer International Publishing Switzerland 2015   177
R.C. Whisonant, *Arming the Confederacy*, DOI 10.1007/978-3-319-14508-2_15

propellant in warfare. Firearms in the United States using old-fashioned black gunpowder last saw action in the Spanish-American War in 1898–1900.

Interest in cave saltpeter, however, did not disappear immediately. The United States Department of Agriculture looked at these deposits for use as fertilizers in the mid-1870s. In 1875 and 1876, niter taken from caverns in Alabama apparently went for this purpose. During World War I studies of saltpeter deposits in more than 100 caves in eastern Tennessee examined the possibility of using the nitrate in the military campaigns raging in Europe. Nevertheless, readily available nitrates from South America and improvements in making synthetics rendered recovery from American cave sediments impractical. Although cave nitrates are no longer mined for raw material, saltpeter-based black powder has never gone away entirely. It still finds application in specialized mine blasting, fireworks, and by hunters, target shooters, and battle re-enactors using classic firearms.

The post-war fates of the Augusta Powder Works and its innovative chief, Brigadier General George W. Rains, provide a fitting epitaph for the end of the Civil War glory days of black powder manufacture. The city of Augusta escaped capture and devastation by the Union, allowing its giant powder plant to keep running after Appomattox until April 18 when Rains shut it down. Later expansion of the city brought down almost all of the factories, foundries, and machine shops, save for one tall chimney made into a monument. Rains stayed in Augusta and became a professor of chemistry at the Medical College of Georgia. Talking to a local veterans group on one occasion, he recalled:

> Here was once heard the noise of the clanking wheels and muffled sounds of the ponderous rollers of war, as they slowly concentrated into black masses the enormous energies which were to shake the earth and air, with the roar and deafening explosions of the battlefield… Of the extensive Confederate Powder Works nothing remains except the obelisk enclosing the great Chimney. Its battlemented tower and lofty shaft, large proportions and beautiful workmanship, will bear evidence of the magnitude and style of their construction to future generations [187].

Rains died in New York in 1898, leaving behind instructions that the Confederate garrison flag he had personally hauled down in April 1865 be buried at the obelisk. For many years, members of the Augusta Chapter of the United Daughters of the Confederacy placed a flag and flowers at the monument's base in his honor.

Lead

As for Wythe County lead, Stoneman's raid in April 1865 demolished many of the facilities at the mines and smelters, but work started up again once the fighting ended. After 1870, the manufacture of lead diminished as fresh discoveries of zinc-rich rocks at the site pushed output of that metal to the forefront. In 1902, with the extraction of zinc now dominant, the New Jersey Zinc Company bought the properties and commenced expanding the installations significantly, turning out more zinc and less lead as the years went by. Caught between low base metal prices and high operating costs, the corporation closed down the Wythe County mines permanently in 1981, sealing the underground chambers and allowing them to flood. This ended 225 years of activity at the Austinville site, at that time a record for the oldest continuously operating mine in the United States.

Not long after the Civil War ceased, the days of a simple naked lead bullet being the main rifle projectile drew to an end. A Swiss army officer in 1882 invented the copper jacketed bullet, a breakthrough that found rapid adoption in global armies along with the high velocity smokeless gun powders then under development as well. These more powerful propellants deformed the traditional soft lead bullets too much, even to the point of melting them slightly. The result was unacceptably poor aerodynamic performance of the distorted projectile and increased lead residues in gun barrels. Encasing a lead core in a stronger metal such as copper or steel solved the problems, enabling startling advances in the evolution of rapid fire, magazine-fed rifles and machine guns. By the advent of World War I, the modern metal ammunition cartridge had taken over as the standard bullet in warfare.

Lead currently has far more uses than as rounds in firearms. Most goes into lead-acid batteries; other applications are in ammunition, glass and ceramics, casting metals, and sheet lead. Since the early eighteenth century start of mining in the immense Mississippi Valley lead deposits, America has grown into a prime supplier of lead in the international markets. Currently rated third behind Australia and China, the country's lead continues to come mainly from the Tri-State District. Missouri in particular accounts for over 90 per cent of the national yield.

Today, most Americans remain unaware of the fundamental role played by the Wythe County mines in the epic confrontation between North and South. All of the old buildings and production facilities at Austinville came down when the mines closed, and no monuments have been erected on the site. The Shot Tower somehow survived unscathed by the fighting and destruction that stormed nearby, serving today as a mute reminder of the remarkable history of the lead operations. (This unusually well preserved structure, one of a handful of shot towers left standing in the country, is now part of the Virginia State Park system and well worth a visit.) Perhaps the strongest links remaining between the Civil War period and the historic lead works are the descendants of the miners themselves, some of whom live in the area and keep alive the memories of their ancestors who provided ammunition for the Confederate armies of 1861–1865.

Salt

Saltville went on making the white crystals after the conflict. With lessened military demands, output at first dropped back nearer to that of pre-war levels, then climbed steadily in the next decades. Stuart, Buchanan and Company, wartime operators of the industry, struggled for a few years after 1865 before reorganizing into the Holston Salt & Plaster Company; that firm stayed in business for the next quarter century. A profound change occurred in 1892 when the Mathieson Alkali Works, a leading chemical manufacturer, acquired the salt operations. Over the next 80 years, the little community evolved into a classic Appalachian "company town," albeit one based on salt instead of coal.

After purchasing the manufacturing properties, Mathieson got busy immediately transforming Saltville. Construction soared, as modern processing plants, houses for the workers and managers, a hotel, and the Mathieson General Store went up. For years a company-run hospital occupied the second floor above the general store. Mathieson made salt until 1906, after which it switched to turning out salt byproducts, among them baking soda and chlorine compounds. The worst calamity of the Mathieson era was the Christmas Eve 1924 disaster when the "muck dam" impounding a huge reservoir of industrial waste failed. The surge downriver took 19 lives; next day rescuers found among the survivors two little girls crying in a bed on the second floor of a house swept downstream. The Mathieson Company became Olin Mathieson Chemical Corporation in 1954 and opened a hydrazine plant in 1961 that made chlorine-based rocket fuel for the NASA space program. When American astronauts left the first footprints in the lunar dust in 1969, Saltville salt powered the Saturn V rocket that took them there.

In 1970, Olin Mathieson announced its intent to close down the Saltville plant, and two years later left town completely. Losing an employer of this magnitude was a shock, but in the years since, the community has worked hard to develop other resources. A number of industries have sprung up, including one using the underground salt caverns to store natural gas on its way to market. Portions of the brines pumped from the subsurface to allow room for the gas are used to make salt for commercial sales, thus keeping alive the centuries-old manufacture of halite. The Museum of the Middle Appalachians is an excellent downtown facility that displays fossils of the giant Ice Age animals drawn to the salt licks, artifacts of the Native Americans who later peopled the Valley, and a host of Civil War materials. The town and museum have partnered with Radford University and the National Park Service to protect and preserve the abundant Civil War earthworks still there.

Today the murders of the black soldiers are commemorated each year around the date of the October 2, 1864, battle in a ceremony conducted by descendants of the men in the 5th USCC. That atrocity occurred with only a few months left in the hostilities. Amid the confusions of a disintegrating government, the accused perpetrators of the executions escaped retribution, one permanently, one for only a short while. General John Breckinridge, incensed at the crimes committed by men under his charge, followed up immediately, telling Lee of the events and charging Champ Ferguson's commander, General Felix Robertson, with complicity in them. Lee ordered Robertson's arrest, but the officer had since departed to Georgia where he fell wounded in action, thereby avoiding an investigation of the killings in Virginia. Robertson never went on trial, instead living peacefully until his death in 1928. Many incorrectly cite him as the last Southern general to die; however, the Confederate Senate did not officially ratify his nomination to that rank, probably owing to the slaughter perpetrated by his troops at Saltville.

Champ Ferguson failed to share the good fortune of the officer under whom he fought in October 1864. Not long after the battle, Breckinridge tracked him down and imprisoned him in Wytheville. Just before the surrender at Appomattox, the Confederate partisan was set free, apparently because no witnesses came forward to testify against him. He returned to his farm in Kentucky and killed two more people, whereupon Federal authorities placed him under arrest. At his trial in Nashville in May 1865, a military court charged Ferguson with the slaying of Lieutenant Elza Smith, 12 other white officers, and two black troopers. In the end, he received the death sentence for killing 53 men and being a "...border rebel guerilla, robber, and murderer" [133]. Men from the 15th Colored Infantry silently witnessed the hanging that next October. Ferguson and Henry Wirz, commander of the infamous Andersonville prisoner of war camp, were the only two Confederates executed as war criminals.

Iron

Western Virginia's iron furnaces and forges, like the other mineral works in spring 1865, lay in disarray and the economic challenges of the following years held revitalization down for a time. As the 1880s arrived, iron making flourished once again, along with increased coal mining from the untapped fields in Southwest Virginia and the rapid expansion of railroads. Improved technology in the form of coke-fired "hot blast" furnaces became widely adopted and profoundly impacted the rebirth of iron manufacture. The progressive techniques, together with the tremendous size of the Valley and Ridge iron deposits, lower labor and operating costs, and proximity to the high-demand Eastern markets, propelled the industry into its "golden age" when Virginia iron gained prominence at the national level.

The state led the country in the output of limonite ore from 1890 to 1901. Fueled by the surge in mining, Virginia rose as high as fourth among all iron-producing states in 1896, 1900, and 1902, and ranked consistently in the top ten from 1889 to 1921. In the revived New River-Cripple Creek District, a dozen charcoal furnaces and forges operated in 1880. Later six or seven big coke smelters appeared, and in 1905 two trainloads of ore were mined and shipped every day from the Cripple Creek area. Virginia's iron enterprises declined precipitously in the 1920s due primarily to competition from cheaper and higher quality Great Lakes ore. All iron making finally closed down around 1930 owing to economic problems rather than exhaustion of the ore reserves.

The only use of the limonite ores anywhere in Virginia today is from the old New River-Cripple Creek District. In southern Pulaski County Hoover Color Corporation extracts iron oxides in open pits for ocher, umber, and sienna coloring pigments. The company is the lone provider of the color sienna in the United States. The last vestige of Virginia's historic iron resources that made Confederate machines of battle currently turns out coloring agents for paints, building materials, and art supplies, including children's crayons.

And what of Tredegar and Joseph Anderson, the soldier-iron maker who built the business into the industrial workhorse of the Southern war effort? Anderson had managed to save his firm from destruction when Richmond fell and then secured pardons for himself and his partners from President Andrew Johnson to resume operations. Much rebuilding had to be done, for the equipment was dilapidated and worse, money was tight. Nonetheless, the far-sighted Anderson had earlier created a special London bank account to set aside funds gained from wartime blockade running of goods to and from Europe, including cotton. This plus money from Northern capitalists got Tredegar running again and the company reorganized in 1867 with Anderson as president.

The late nineteenth century brought times of hardship for the Tredegar Iron Works. The panic of 1873 and the depression afterwards put it in receivership in 1876. Anderson saved his firm once again but problems persisted. Breakthroughs in steel manufacturing made that metal much better than iron in the new skyscrapers, military hardware, and a host of other applications springing up in the accelerating American economy. Tredegar could not afford to switch to steel fabrication, and steel makers in places like Pittsburgh, Pennsylvania, and Birmingham, Alabama, easily outcompeted the Richmond mills. Yet the business kept going after Anderson's death, turning out specialized iron products that included shells and projectiles for the United States Army and Navy in the country's next four wars—the Spanish-American, World Wars I and II, and Korea. In 1987, descendants of Anderson sold the last of the equipment and the doors of the Confederacy's chief heavy weapons provider closed forever.

Joseph Anderson died in New Hampshire while vacationing in 1892 and his body came back to the Virginia capital for internment in Hollywood Cemetery, resting place of many Confederate notables including Jefferson Davis. The gravesite overlooks the James River and Anderson's beloved iron factories along the water's edge. Recent preservation efforts have restored parts of Tredegar, notably the gun foundry and its imposing 86-foot-tall chimney. A first-rate museum recounts the fundamental part the iron works played in the nation's industrial growth and in the Civil War, all of this a legacy of the remarkably resourceful ironmaster of the Confederacy.

Coal

Most of the money earned by today's Virginia minerals industry is from coal, and has been for decades. The state presently ranks twelfth among the 26 coal-generating states. Even so, none of this fuel is taken from the historic mining centers in the Richmond Basin and the Valley coal fields. Instead, all active operations reside in the distant southwestern corner of the state, known locally as the Pocahontas fields for the main underground seam. This coal-rich area and the vast measures in adjacent West Virginia were first opened a few years after the Civil War, and quickly rose to dominate the older producing sites.

In the Richmond vicinity, the Midlothian mines maintained operation after the war concluded, though constantly plagued by fires, explosions, and worsening water problems. Troubles like these plus rising competition caused the number of mines to drop off sharply; an 1880 United States government map depicted only four mines active in the entire locale. Midlothian mining struggled on into the twentieth century, but the end soon arrived due to the old problems and exhaustion of the economically viable seams. Unable to overcome these difficulties, the last mines shut down in 1927.

At present the Richmond coal measures are much changed from past times when numerous mines humming with activity dotted the landscape. Pleasant and prosperous, several of the previous mining villages, most especially Midlothian itself, live on as suburban commuter communities. The dense network of coal railroads is gone and the mine entrances or pits are mostly vegetated over and hard to find. The workmen quit swinging their digging tools decades ago, yet on the official seal of Chesterfield County, center of the Midlothian fields, a miner leans on a large coal pickaxe, a reminder of the region's key role in the nation's economic development and the South's bid for independence.

Across the state to the west, coal extraction in the Valley fields resumed after the fighting stopped; however, tonnage stayed low for the rest of the nineteenth century. All of the mines remained far from the main railroad line (by then, the Norfolk and Western) and this seriously limited growth of the industry. Marketing efforts received another setback when the Norfolk and Western built a railroad to the fast-rising Pocahontas fields farther southwest in 1883. This bituminous coal was much more plentiful, easier to dig, and now readily transportable to the big eastern United States buyers, causing a swift drop in Valley semi-anthracite sales.

In the first decade of the twentieth century, demand for the Valley fields coal soared once again. A widespread strike by Pennsylvania anthracite miners played a part in this, and completion of the long-sought railroad system within the Montgomery County measures helped boost yield also. The expanded tracks connected with the Norfolk and Western's main line, putting Merrimac production into the cars of the foremost coal-hauling railroad in West Virginia and southwestern Virginia. This plus an additional new line in 1907 carried Montgomery County coal to Hampton Roads, a significant export center for domestic and foreign markets. Spurred by these advancements, the semi-anthracite output throughout the Valley fields continued to increase. A disastrous flood from heavy rains closed the Merrimac shafts in 1909, but the mine reopened in 1918 and reached its peak activity in the 1920s and early 1930s. Following some labor unrest, the Merrimac mine closed again in 1935, this time permanently. Coal mining persevered elsewhere in the Valley fields until the late 1950s when it ceased entirely save for a few small truck mines that ran sporadically in the 1970s and 1980s.

There is no coal extraction anywhere in the Valley fields today. In Montgomery County, the Coal Mining Heritage Park and Loop Trail serves to commemorate the 200-year-long history of the county mines. At Merrimac, a few ruins recall that location's proud past. One last project in the late 1980s reclaimed limited amounts of coal from waste piles at the old Merrimac site. This material was cleaned in a shaker and sent to a North Carolina charcoal briquette company for final processing and sales. From the bunkers of the man-of-war *Virginia*, perhaps, to backyard barbecues—thus slipped the heritage of the Confederate coal mine in Montgomery County into the mists of history.

The Virginia and Tennessee Railroad

In the Civil War, six separate raids targeted the Virginia and Tennessee Railroad, including the May 1864 attack on Dublin and the New River Bridge at Central Depot. The devastation inflicted was severe; the company's annual report on June 30, 1865, recorded that only three bridges and three depots escaped the fighting unharmed. Despite the pervasive damage, however, the trains never stopped running completely. Shortly after peace returned, a Southern veteran from Pulaski County, released from prisoner of war detainment in Maryland, wrote about making his way back home in July 1865. He traveled the last leg of the trip on the Virginia and Tennessee all the way to the station at Dublin. With rebuilding of the rail line well under way in 1867, former Confederate General William Mahone took over the business and became president. Mahone already controlled two other Virginia rail lines, the South Side Railroad and Norfolk and Petersburg, and aimed to combine them with his latest acquisition. This he successfully accomplished in 1870, merging all three into the Atlantic, Mississippi and Ohio Railroad.

Mahone's rail network prospered for a few years, but the financial Panic of 1873 brought serious problems to this system and many others in the reconstructing South. The Atlantic, Mississippi and Ohio limped on through receivership until the general's decade of management ended in 1881. That year a Philadelphia banking firm purchased the railroad, reorganizing and renaming it the Norfolk and Western. It was this corporation that expanded into the prolific southwestern Virginia and West Virginia coal fields, thereby commencing the boom years of hauling coal, iron, and timber in the late nineteenth century. In the early years of the twentieth century, smaller companies constructed spur lines from the Norfolk and Western main line into the Valley fields that revitalized the historic Merrimac mines and others; the growing railroad giant eventually also acquired those tracks.

In the twentieth century, the Norfolk and Western emerged as one of the premier coal-carrying lines in the nation. In 1982 the railroad merged with the Southern Railway to form the Norfolk Southern Railway. Coal from six major mining states in the East, including Virginia, continues to be the principal commodity transported by its trains. Much of the old Virginia and Tennessee route lives on as part of the Norfolk Southern properties. The tracks cross the New River on a span in exactly the same place as they did in the Civil War, and some of the stonework in the modern bridge pilings probably witnessed the cannon's violent explosions that glorious spring day in May 1864.

The May 1864 New River Valley Railroad Raid

The Battle of Cloyds Mountain and next day's artillery duel at the New River Bridge involved an astonishing number of men who went on to postbellum achievement and fame. Indeed, the 23rd Ohio that smashed through the Confederate lines at Cloyds Mountain to precipitate the Union victory has been called "…the most remarkable regiment of either army, North or South" [139]. At one time or another in the hostilities, future luminaries including two presidents, one of the nation's most highly regarded Indian fighters, a United States Senator and Supreme Court Justice, various other politicians, and even an owner and publisher of the *Los Angeles Times* served in the 23rd.

Colonel Rutherford B. Hayes, an early commander of the 23rd and leader of its brigade at Cloyds Mountain, finished his distinguished Civil War career as a brevet Major General. Elected to serve in Congress before he mustered out of uniform, he rose swiftly in Republican circles and won the presidency in an extremely contentious 1876 election. Hayes was always a man of integrity and did much to restore honor to his high office after the scandal-plagued, post-Civil War administrations of the later nineteenth century. The Ohio officer consistently treasured his military service above all other accomplishments; he preferred to be called "Colonel" until he died in 1893.

Like Hayes, Lieutenant William McKinley saw action at Cloyds Mountain and New River Bridge, compiled an excellent service record, and achieved the highest post-war success in politics. His prominent career as a Republican politician received considerable help from his old friend and comrade-in-arms, Rutherford B. Hayes. In something of a surprise, McKinley gained his party's nomination and the presidency in 1896. Among the most important events in his first term was the Spanish-American War, which he entered reluctantly having experienced firsthand the horrors of battlefield slaughter at places like Cloyds Mountain. His tragic assassination in Buffalo in 1901 ended the long line of Union military heroes elected president of the United States.

Following the Virginia and Tennessee Railroad raid in 1864, General George Crook participated in the Shenandoah Valley campaigns and the last battles at Richmond and Appomattox. Afterwards he stayed in the service, ultimately becoming acclaimed by General Sherman as "...the greatest Indian fighter...the United States Army ever had" [200]. President Grant gave him command of the Arizona Territory in 1872, after which he went north to fight in the Sioux Wars of 1876–1877. During that time Crook's unit made up part of a larger offensive against the Sioux and Cheyenne that resulted in General George Custer's annihilation at the Little Big Horn in 1876. In the 1880s, the general gained national recognition as the man who captured Geronimo, although it was more a case of talking the fabled Apache chief into surrendering. Crook spent much time in his later years striving for fair treatment of the native peoples for whom he had much respect. Upon his death in 1890 a Sioux chieftain commented "...he, at least, had never lied to us. His words gave the people hope" [201].

On the Confederate side, Colonel John McCausland, whose skillful retreat from the Cloyds Mountain battlefield saved his small force, received promotion to brigadier general a few days after that fight and his spirited defense of the New River Bridge. In summer 1864 he served under General Jubal Early in the raid north from the Shenandoah Valley into Maryland and Pennsylvania. On July 30 Early had him burn Chambersburg, Pennsylvania, in retribution for previous Union devastations in the Virginia Valley, particularly the torching of the Virginia Military Institute. (McCausland graduated from VMI and taught there for two years in the 1850s.)

Later in the struggle, McCausland fought under Lee at Richmond and Appomattox Court House. When the conflict ended, McCausland, fearing arrest by Federal authorities for razing Chambersburg, fled to Europe and Mexico for two years. Pardoned by President Grant whom he had befriended in St. Louis before the war, McCausland returned home and lived quietly on his farm in West Virginia until his death in 1927. Unreconstructed to the end, the gritty old warrior was in fact the last Rebel general to die.

The career of Confederate Sergeant Milton Humphreys during and after the war is perhaps the strangest of all those who contested at Cloyds Mountain and New River Bridge. As a youth in western Virginia, Humphreys showed unusual academic abilities, especially in mathematics. When the Civil War broke out, he gave up his studies at Washington College (now Washington and Lee University) in Lexington, Virginia, and enlisted in Bryan's Battery of the 13th Virginia Light Artillery. In May 1862, Humphrey's unit was campaigning in West Virginia when he got a chance to test his idea of indirect fire, or shooting at an unseen target. This had rarely been done because estimating the distance to the target involved mostly guesswork. Using his skills in trigonometry, the 18-year-old gunnery sergeant calculated the muzzle elevation needed to drop a shell onto a Union position hidden by trees. It succeeded, and the confused enemy soldiers had no idea where the shots raining down on them came from.

After peace returned, Humphreys finished his undergraduate education and earned a doctorate from the University of Leipzig in what would become the German state of Saxony. He later taught at Vanderbilt University in Nashville, Tennessee, the University of Texas in Austin, and finally at the University of Virginia in Charlottesville from which he retired in 1912. He died in 1928 and was buried in the University chapel. Like Hayes and McKinley, his opponents in the 1864 New River Valley campaign, the renowned academician prized his military service above all achievements and honors. Humphreys once commented that serving as First Gunner in Bryan's Battery was "…a title in which I take more pride than any other ever bestowed upon me" [38]. He never claimed credit for inventing the technique of indirect fire because "…the thing is so obvious" [38]. Yet if he did, this exceptional non-commissioned officer went on to say, the concept did not originate that day in battle in 1862 but "…in my day dreams when I was about 8 years old" [38].

The mineral works and railroad in southwestern Virginia were defended to the bitter end and beyond. On April 10, 1865, one day after Lee's surrender, word of the event reached the 600 or so soldiers left in the region of the 10,000 who had fought so tenaciously against the Northern invasions. With this news, their officers put to a vote whether to go home peacefully or continue to bear arms. Ten men elected to stack their weapons while the remainder chose to go on resisting. Two days later a small engagement, perhaps the last combat that occurred in Virginia, took place near Central Depot when Federal troops, once again attacking the railroad, encountered a little band of the Confederate holdouts. In a brief skirmish known locally as the Battle of Seven Mile Tree, the Union men were driven away. The Southerners afterwards retreated into North Carolina and disappeared from the pages of history.

In these final days of fighting, General John Breckinridge had once again assumed command of the Department of Southwest Virginia and East Tennessee. In his previous stint in late 1864 as the department's leader, Breckinridge had directed his men well throughout the defense of the salt works, lead mines, and the Virginia and Tennessee Railroad. Very conscious of conducting the war with integrity, he felt genuine outrage at the massacre in Saltville and urged Lee to do something about it. Appointed to serve as Confederate Secretary of War in January 1865, he soon saw that the contest could not be won and discussed ending the hostilities honorably with Jefferson Davis. On May 8, almost a month after Appomattox, it fell to Breckinridge to officially disband his department. Gathering his few veterans before him, he thanked them and sent them away saying "You have done all that can be done...for us, the long agony is over" [245]. Perhaps these words befit all the soldiers, North and South, who contended in the bloody campaigns and battles in the ancient mountains of southwestern Virginia.

Notes

The Epilogue presents an account of the fate of the mineral operations and the Virginia and Tennessee Railroad after the war. The post-war activities of some of the people involved in the May 1864 railroad raid are also described. Many of the sources used in the individual chapters dealing with these topics were referenced again for the Epilogue, but some additional works provided information also.

Specifically, the niter section includes information from Rains [188], Sipe [214], Melton [150], De Paepe and Hill [44], and Kelly [105]. For the discussion of lead, the post-war history of the Wythe County lead mines comes from Weinberg [255, 256]; the development of modern lead bullets from Moore [157] and Wallace [249]; and the current status of lead production in the United States from the U.S. Geological Survey 2014 website [232]. In the salt part of the Epilogue, the development of the operations in Saltville after the war is summarized from Kent [110], Sarvis [199], Allison [1], and the Saltville Historical Society undated. The follow-up of the black soldiers massacre, including the arrest and hanging of Champ Ferguson, is from Marvel [129] and Mays [134, 136]. The information about the post-1865 resurgence of the iron industry in western Virginia and its ultimate decline is from Gooch [69], save for the brief account from Sweet and Nolde [229] of the last remaining use of the iron ore as a coloring agent. For the eventual fate of Joseph Anderson and the Tredegar Works I used Crews [37] primarily, supplemented by Bruce [28], Norville [171], and Schult [206]. My discussion of the late nineteenth and twentieth century coal mining operations is compiled from Weaver [254], Wilkes [267], and Hibbard [80] for the Richmond fields, and Hibbard [80], Proco [184], and La Lone [116] for the Montgomery County works. The evolution of the Virginia and Tennessee Railroad since 1865 is abstracted from Smith [215], Lambie [117], Noe [169], and the Norfolk Southern 2014 website [170]. The section discussing the postbellum careers of some combatants in the May 1864 railroad raid is from the following sources: McLean [140] for the 23rd Ohio Volunteer Infantry regiment, Rutherford B. Hayes (supplemented by Williams [268]), William McKinley, and John McCausland; Schmitt [202] for George Crook; and Crookshanks [39] for Milton Humphreys. The Crookshanks article was originally published in *America's Civil War* magazine and is posted online. The concluding paragraphs on the end of the war in southwestern Virginia are from Walker [246].

Bibliography

1. Allison RA (1996) A brief history of Saltville. Saltville Centennial Committee, Saltville, VA
2. Anderson RV (1996) Forging iron and slavery in Valley of Virginia. The Washington Times, Saturday, November 30, 1996, p B3
3. Anonymous (1857) A winter in the South. Harper's New Monthly Magazine 15(88), September 1857, p 448
4. Anonymous (1857) A winter in the South. Harper's New Monthly Magazine 15(88), September 1857, p 433–451
5. Anonymous (1993) Shot tower historical State Park, Wythe County, Virginia. Department of Conservation and Recreation, Richmond, VA
6. Armstrong JT (1986) History of Smyth County, Virginia, volume Two, 1832–1870: ante-bellum years through the Civil War. Smyth County Historical and Museum Society, Inc., Marion, VA, p 131
7. Armstrong JT (1986) History of Smyth County, Virginia, Volume Two, 1832–1870: ante-bellum years through the Civil War. Smyth County Historical and Museum Society, Inc., Marion, VA
8. Arthur EC (1889) The Dublin raid. Ohio Soldier 2, January 5–April 13, 1889, p 386
9. Arthur EC (1889) The Dublin raid. Ohio Soldier 2, January 5–April 13, 1889, pp 386–387
10. Asimov I (1981) In the beginning: science faces god in the book of genesis. Stonesong Press, LLC, New York
11. Atack J, Passel P (1994) A new economic view of American history. W. W. Norton & Company, New York
12. Austin VL (1977) The Southwest Virginia lead works, 1756–1802. M.A. Thesis, Virginia Polytechnic Institute and State University, Blacksburg
13. Bartholomew MJ, Brown KE (1992) The Valley Coalfield (Mississippian age) in Montgomery and Pulaski counties, Virginia. Virginia Division of Mineral Resources Publication 124
14. Bartholomew MJ, Mills HH (1991) Old courses of the New River: its late Cenozoic migration and bedrock control inferred from high-level stream gravels, southwestern Virginia. Geol Soc Am Bull 103:73–81
15. Black RC III (1952) The railroads of the Confederacy. University of North Carolina Press, Chapel Hill, p 243
16. Black RC III (1952) The railroads of the Confederacy. University of North Carolina Press, Chapel Hill, p 200
17. Black RC III (1952) The railroads of the Confederacy. University of North Carolina Press, Chapel Hill, p 191
18. Black RC III (1952) The railroads of the Confederacy. University of North Carolina Press, Chapel Hill

© Springer International Publishing Switzerland 2015
R.C. Whisonant, *Arming the Confederacy*, DOI 10.1007/978-3-319-14508-2

19. Bocian M, Salmon J (2012) The Virginia central railroad during the Civil War. Encyclopedia of Virginia. http://www.EncyclopediaVirginia.org/Virginia_Central_Railroad_During_the_Civil_War_The

20. Boyd JP (ed) (1952) The papers of Thomas Jefferson, vol 6, 21 May 1781 to 1 March 1784. Princeton University Press, Princeton, p 201

21. Boyd JP (ed) (1952) The papers of Thomas Jefferson, vol 6, 21 May 1781 to 1 March 1784. Princeton University Press, Princeton

22. Boyle RS (1936) Virginia's mineral contribution to the Confederacy. Virginia Division of Mineral Resources Bulletin 46:119–123

23. Brady TT (1991) The charcoal iron industry in Virginia. Virginia Division of Mineral Resources. Virginia Miner 37(4):27–31

24. Brown RH (1948) Historical geography of the United States. Harcourt, Brace and Company, New York, p 130

25. Brown RH (1948) Historical geography of the United States. Harcourt, Brace and Company, New York

26. Brown A (1962) Geology and the Gettysburg campaign, vol 5, Educational series. Pennsylvania Geological Survey, Harrisburg

27. Bruce K (1930) Virginia iron manufacture in the slave era. The Century Company, New York, p 407

28. Bruce K (1930) Virginia iron manufacture in the slave era. The Century Company, New York

29. Burkhart OC (1931) County mining history traced. Blacksburg News Messenger, Wednesday, May 6, 1931, p 1

30. Cato KD, Gelinas R, Kemppinen H, Amick D (1994) The genesis of Mountain Lake: a landslide dam origin. Association of Engineering Geologists, 37th Annual Meeting, Williamsburg, VA, Program and Abstracts, p 42

31. Coggins J (1962) Arms and equipment of the Civil War. Doubleday and Company, Inc., New York, p 26

32. Coggins J (1962) Arms and equipment of the Civil War. Doubleday and Company, Inc., New York

33. Cooper BN (1966) Geology of the salt and gypsum deposits in the Saltville area, Smyth and Washington Counties, Virginia. In: Rau JL (ed) Second symposium on salt, vol 1, Geology, geochemistry, and mining. Northern Ohio Geological Society, Cleveland, pp 11–34, figures 1–12

34. Corrosion Doctors (2012) http://corrosion-doctors.org/Elements-Toxic/Lead-history.htm

35. Cowen R (1999) Essays on geology, history, and people. http://mygeologypage.ucdavis.edu/cowen/~GEL115

36. Craig JR, Vaughn DJ, Skinner BJ (2001) Resources of the earth; origin, use, and environmental impact. Prentice Hall, Upper Saddle River, NJ

37. Crews ER (1992) The industrial bulwark of the Confederacy. Am Herit Invent Technol 7(3):8–17

38. Crookshanks B (2006) p 3. http://www.historynet.com/sergeant-milton-humphreys-concept-of-indirect-fire.htm

39. Crookshanks B (2006) http://www.historynet.com/sergeant-milton-humphreys-concept-of-indirect-fire.htm

40. Daniel JRV (1951) Jack Jouett and Paul Revere in petticoats: the heroine of the battle of Wytheville. Virginia Cavalcade 1(1):35

41. Daniel JRV (1951) Jack Jouett and Paul Revere in petticoats: the heroine of the battle of Wytheville. Virginia Cavalcade 1(1):33–36

42. Davis WC (1971) The massacre at Saltville. Civ War Times Illus 9:4–11, 43–48

43. De Paepe D (1981) Saltpeter mining features and techniques. National Speleological Society Bulletin 43:103–105

44. De Paepe D, Hill CA (1981) Historical geography of United States saltpeter caves. National Speleological Society Bulletin 43:88–93

45. Dietrich RV (1970) Geology and Virginia. Virginia Division of Mineral Resources, Charlottesville

46. Donnelly RW (1959) The Confederate lead mines of Wythe County, Va. Civil War Hist 5(4):403
47. Donnelly RW (1959) The Confederate lead mines of Wythe County, Va. Civ War Hist 5(4):402–414
48. Dowdey C (1955) The land they fought for; the story of the south as the Confederacy. Doubleday, Garden City, NY
49. Egan M (1888) The flying, gray-haired yank: or, the adventures of a volunteer. Hubbard Brothers, Philadelphia, PA, p 167
50. Egan M (1888) The flying, gray-haired yank: or, the adventures of a volunteer. Hubbard Brothers, Philadelphia, PA
51. Emerson F (1996) The battle of the Cove. In: Hoch BR (ed) Wythe County Virginia during the war between the states 1861–1865. The Town of Wytheville and the Wythe County Historical Society, Wytheville, p 31
52. Emerson F (1996) The battle of the Cove. In: Hoch BR (ed) Wythe County Virginia during the war between the states 1861–1865. The Town of Wytheville and the Wythe County Historical Society, Wytheville, p 32
53. Emerson F (1996) The battle of the Cove. In: Hoch BR (ed) Wythe County Virginia during the war between the states 1861–1865. The Town of Wytheville and the Wythe County Historical Society, Wytheville, pp 23–32
54. Emerson F (1996) Yankee raids and skirmishes. In: Hoch BR (ed) Wythe County Virginia during the war between the states 1861–1865. The Town of Wytheville and the Wythe County Historical Society, Wytheville, pp 33–38
55. Evans D (1993) Stoneman's raids. In: Current RN (ed) Encyclopedia of the Confederacy. Simon and Schuster, New York, p 1547
56. Evans D (1993) Stoneman's raids. In: Current RN (ed) Encyclopedia of the Confederacy. Simon and Schuster, New York, pp 1546–1547
57. Faust B (1960) The last of the petre monkeys. National Speleological Society News 17:11
58. Faust B (1960) The last of the petre monkeys. National Speleological Society News 17:10–12
59. Faust B (1964) Saltpetre caves and Virginia history. In: Douglas HH (ed) Caves of Virginia. Virginia Cave Survey, Falls Church, VA, p 32
60. Faust B (1964) Saltpetre caves and Virginia history. In: Douglas HH (ed) Caves of Virginia. Virginia Cave Survey, Falls Church, VA, p 33
61. Faust B (1964) Saltpetre caves and Virginia history. In: Douglas HH (ed) Caves of Virginia. Virginia Cave Survey, Falls Church, VA, p 47
62. Faust B (1964) Saltpetre caves and Virginia history. In: Douglas HH (ed) Caves of Virginia. Virginia Cave Survey, Falls Church, VA, p 49
63. Faust B (1964) Saltpetre caves and Virginia history. In: Douglas HH (ed) Caves of Virginia. Virginia Cave Survey, Falls Church, VA, pp 31–56
64. Fordney BE (1999) Personality: George Stoneman led the Army of the Potomac's first cavalry corps and later served as governor of California. Military Hist April 1999, p 18, 20, 80
65. Freeman J, Crook E, Watts C (2012) Seismic refraction survey of landslide colluvium and lacustrine sediments at Mountain Lake, Virginia. Geological Society of America Annual Meeting, Charlotte, NC, Abstracts with Programs, vol 44, n 7, p 390
66. Freis R (1996) Price Mountain's place in history is significant. Roanoke Times, Sunday, March 3, 1996, New River Current Section, pp 20–21
67. Freis R (1998) Unbroken circle: coal mining in Montgomery County. Virginia Polytechnic and State University, Blacksburg, VA, Unpublished manuscript
68. Frye K (1986) Roadside geology of Virginia. Mountain Press Publishing Company, Missoula, MT
69. Gooch EO (1954) Iron in Virginia. Virginia Division of Mineral Resources, Mineral Resources Circular 1, 17 p
70. Gray MP (2011) Manassas gap railroad during the Civil War. Encyclopedia Virginia. http://www.EncyclopediaVirginia.org/Manassas_Gap_Railroad_During_the_Civil_War
71. Guerrant EO (1999) Bluegrass Confederate: the headquarters diary of Edward O. Guerrant, edited by W. C. Davis and M. L. Swentor. Louisiana State University Press, Baton Rouge, p 546

72. Guerrant EO (1999) Bluegrass Confederate: the headquarters diary of Edward O. Guerrant, edited by W. C. Davis and M. L. Swentor. Louisiana State University Press, Baton Rouge, p 547

73. Guerrant EO (1999) Bluegrass Confederate: the headquarters diary of Edward O. Guerrant, edited by W. C. Davis and M. L. Swentor. Louisiana State University Press, Baton Rouge, p 613

74. Guerrant EO (1999) Bluegrass Confederate: the headquarters diary of Edward O. Guerrant, edited by W. C. Davis and M. L. Swentor. Louisiana State University Press, Baton Rouge

75. Hamblin WK, Christiansen EH (2003) Earth's dynamic systems, 10th edn. Prentice Hall, Upper Saddle River, NJ

76. Hauer PM (1970) Spelean time and nitre. National Speleological Society News 28:85–86

77. Hawley D (2011) Engineering the Union's victory. The History Channel Magazine, July/August 2011, p 33

78. Hawley D (2011) Engineering the Union's victory. The History Channel Magazine, July/August 2011, pp 32–38

79. Hibbard WR Jr (1990) Virginia coal: an abridged history. Virginia Center for Coal and Energy Research, Blacksburg, p 17

80. Hibbard WR Jr (1990) Virginia coal: an abridged history. Virginia Center for Coal and Energy Research, Blacksburg

81. Hibbard WR Jr (1993) Mining. In: Current RN (ed) Encyclopedia of the Confederacy. Simon and Schuster, New York, pp 1043–1044

82. Hill CA (1981) Origin of cave saltpeter. National Speleological Bulletin 43:110–126

83. Holden RJ (1907) Iron. In: Watson TL (ed) Mineral resources of Virginia. J. P. Bell Co., Lynchburg, VA, pp 402–491

84. Holmes ME (1993) Salt. In: Current RN (ed) Encyclopedia of the Confederacy. Simon and Schuster, New York, p 1363

85. Holmes ME (1993) Salt. In: Current RN (ed) Encyclopedia of the Confederacy. Simon and Schuster, New York, pp 1362–1363

86. Houser BB (1981) Erosional history of the New River, southern Appalachians, Virginia. U. S. Geological Survey Open-File Report 81–771

87. Hoyle SJ (1997) Mining in the Civil War: the skirmish line. Newsletter of the Civil War Round Table of Southern West Virginia 1(3):1–8

88. Hubbard DA Jr (1996) A Virginia classic: Clarks cave. The Virginia Cellars 5:3–6, 33–34

89. Hubbard DA Jr, Mitchell RS, Herman JS (1986) The mineral and chemical constituents of saltpeter in six Virginia caves. Communications of the 9th Congress Internacional de Espeleologia, Spain, vol 2, pp 67–70

90. Hudson JC (2002) Across this land: a regional geography of the United States. Johns Hopkins University Press, Baltimore, MD, p 144

91. Hudson JC (2002) Across this land: a regional geography of the United States. Johns Hopkins University Press, Baltimore, MD

92. Humphreys MW (1924) A history of the Lynchburg campaign. The Michie Co., Charlottesville, VA, p 19

93. Humphreys MW (1924) A history of the Lynchburg campaign. The Michie Co., Charlottesville, VA, p 27

94. Humphreys MW (1924) A history of the Lynchburg campaign. The Michie Co., Charlottesville, VA

95. Jensen ML, Bateman AM (1981) Economic mineral deposits, 3rd edn. Wiley, New York

96. Johnson PG (1986) The United States Army invades the New River Valley, May 1864. Walpa Publishing, Christiansburg, VA, p 42

97. Johnson PG (1986) The United States Army invades the New River Valley, May 1864. Walpa Publishing, Christiansburg, VA, p 97

98. Johnson PG (1986) The United States Army invades the New River Valley, May 1864. Walpa Publishing, Christiansburg, VA

99. Johnson F (1996) Toland's raid: battle of Wytheville July 18. In: Hoch BR (ed) Wythe County Virginia during the war between the states 1861–1865. The Town of Wytheville and the Wythe County Historical Society, Wytheville, pp 15–22

100. Johnston AJ II (1961) Virginia railroads in the Civil War. University of North Carolina Press, Chapel Hill
101. Kegley MB (1989) Wythe county Virginia a bicentennial history. Wythe County Board of Supervisors, Wytheville, p 331
102. Kegley MB (1989) Wythe county Virginia a bicentennial history. Wythe County Board of Supervisors, Wytheville, p 336
103. Kegley MB (1989) Wythe County Virginia a bicentennial history. Wythe County Board of Supervisors, Wytheville
104. Kegley MB (2003) I like Molly Tynes whether she rode or not. Kegley Books, Wytheville
105. Kelly J (1998) Explosive growth. Invention & Technology, Spring 1998, pp 10–20
106. Kennedy P (1989) The rise and fall of the great powers. Vintage Books, New York, p 439
107. Kennedy P (1989) The rise and fall of the great powers. Vintage Books, New York, p 180
108. Kennedy P (1989) The rise and fall of the great powers. Vintage Books, New York
109. Kent WB (1955) A history of Saltville, Virginia. Commonwealth Press, Radford, VA, p 29
110. Kent WB (1955) A history of Saltville, Virginia. Commonwealth Press, Radford, VA
111. Ketchum RM (ed), Catton B (narrative) (1960) American heritage picture history of the Civil War. American Heritage Publishing Company, Inc., New York, p 396
112. Ketchum RM (ed), Catton B (narrative) (1960) American heritage picture history of the Civil War. American Heritage Publishing Company, Inc., New York
113. Kurlansky M (2003) Salt: a world history. Penguin Books, New York, p 260
114. Kurlansky M (2003) Salt: a world history. Penguin Books, New York, p 222
115. Kurlansky M (2003) Salt: a world history. Penguin Books, New York
116. La Lone MB (1997) Appalachian coal mining memories. Pocahontas Press, Inc., Blacksburg
117. Lambie JT (1954) From mine to market; the history of coal transportation on the Norfolk and western railway. New York University Press, Washington Square, NY
118. Lewis WC (1989) Some historical speculations on the origin of saltpeter. National Speleological Society Bulletin 51:66–70
119. Lincoln A (1863) The emancipation proclamation. United States National Archives & Records Administration, Washington, DC, p 4. http://www.archives.gov/exhibits/featured_documents/emancipation_proclamation/
120. Lincoln A (1863) The emancipation proclamation. United States National Archives & Records Administration, Washington, DC. http://www.archives.gov/exhibits/featured_documents/emancipation_proclamation/
121. Linklater A (2002) Measuring America: how an untamed wilderness shaped the United States and fulfilled the promise of democracy. Walker and Company, New York, p 36
122. Linklater A (2002) Measuring America: how an untamed wilderness shaped the United States and fulfilled the promise of democracy. Walker and Company, New York
123. Lonn E (1933) Salt as a factor in the Confederacy. Walter Neale, New York, p 13
124. Lonn E (1933) Salt as a factor in the Confederacy. Walter Neale, New York
125. Lowry WD (1989) Mississippian age of the New River in the valley and Ridge province of Virginia. Proceedings, New River Symposium, Radford, 20–22 Apr 1989
126. Lynch ME (2001) Confederate war industry: the niter and mining bureau. M.S. Thesis, Virginia Commonwealth University, Richmond, p 57
127. Lynch ME (2001) Confederate war industry: the niter and mining bureau. M.S. Thesis, Virginia Commonwealth University, Richmond
128. Mandigo HM (1986) Manufacturing iron along the New River in Virginia. In: Kegley MB (ed) Glimpses of Wythe County. Central Virginia Newspapers, Inc., Orange, VA, pp 115–124
129. Marvel W (1991) The battle of Saltville: massacre or myth? Blue and Gray magazine August 1991, pp 10–19, 46–60
130. Marvel W (1992) The battles for Saltville. H. E. Howard, Inc., Lynchburg, VA, p 134
131. Marvel W (1992) The battles for Saltville. H. E. Howard, Inc., Lynchburg, VA
132. Mays TD (1998) The Saltville massacre. McWhiney Foundation Press, McMurry University, Abilene, TX, p 13

133. Mays TD (1998) The Saltville massacre. McWhiney Foundation Press, McMurry University, Abilene, TX, p 71
134. Mays TD (1998) The Saltville massacre. McWhiney Foundation Press, McMurry University, Abilene, TX
135. Mays TD (2008) Cumberland blood: champ Ferguson's Civil War. Southern Illinois University Press, Carbondale, p 120
136. Mays TD (2008) Cumberland blood: champ Ferguson's Civil War. Southern Illinois University Press, Carbondale
137. McDonald JN (1984) The Saltville, Virginia, locality: a summary of research and field trip guide. Symposium on the Quaternary of Virginia, Charlottesville
138. McDonald JN, Bartlett CS (1983) An associated Musk Ox from Salville, Virginia. J Paleontol 2(4):453–470
139. McLean GA Jr (2012) Skirmish at Pearisburg. Blackwell Press, Lynchburg, VA, p 3
140. McLean GA Jr (2012) Skirmish at Pearisburg. Blackwell Press, Lynchburg, VA
141. McManus HR (1989) The battle of Cloyds Mountain. H. E. Howard, Inc., Lynchburg, VA, p 26
142. McManus HR (1989) The battle of Cloyds Mountain. H. E. Howard, Inc., Lynchburg, VA, p 72
143. McManus HR (1989) The battle of Cloyds Mountain. H. E. Howard, Inc., Lynchburg, VA, p 78
144. McManus HR (1989) The battle of Cloyds Mountain. H. E. Howard, Inc., Lynchburg, VA
145. McPherson JM (1988) Battle cry of freedom: the Civil War era. Oxford University Press, New York, p 12
146. McPherson JM (1988) Battle cry of freedom: the Civil War era. Oxford University Press, New York, p 675
147. McPherson JM (1988) Battle cry of freedom: the Civil War era. Oxford University Press, New York
148. McPherson JM (2007) This mighty scourge: perspectives on the Civil War. Oxford University Press, New York, p 49
149. McPherson JM (2007) This mighty scourge: perspectives on the Civil War. Oxford University Press, New York
150. Melton M (1973) "A grand assemblage": George W. Rains and the Augusta Powder Works. Civ War Times Illus 11:28–37
151. Miller WJ (1993) Mapping for Stonewall: the Civil War service of Jed Hotchkiss. Elliott and Clark Publishing, Washington, DC
152. Mills HH (1988) Surficial geology and geomorphology of the Mountain Lake area, Giles County, Virginia, including sedimentological studies of colluvium and boulder streams. U. S. Geological Survey Professional Paper 1469
153. Mirsky A (1991) Treasure map: geologic resources of the original thirteen states. Indiana University—Purdue University at Indianapolis (Brochure, 8 p)
154. Mirsky A (1997) Influence of geologic factors on ancient Egyptian civilization. J Geosci Educ 45:415–424
155. Mirsky A, Bland EL (1996) Influence of geologic factors on ancient civilizations in the Aegean area. J Geosci Educ 44:30
156. Mirsky A, Bland EL (1996) Influence of geologic factors on ancient civilizations in the Aegean area. J Geosci Educ 44:25–35
157. Moore W (1963) Guns: the development of firearms, air guns, and cartridges. Grosset and Dunlap, New York
158. Mosgrove GD (1957) Kentucky Cavaliers in Dixie: reminiscences of a Confederate Cavalryman, edited by B. I. Wiley. McCowart-Mercer Press, Jackson, TN, p 202
159. Mosgrove GD (1957) Kentucky Cavaliers in Dixie: reminiscences of a Confederate Cavalryman, edited by B. I. Wiley. McCowart-Mercer Press, Jackson, TN, p 203
160. Mosgrove GD (1957) Kentucky Cavaliers in Dixie: reminiscences of a Confederate Cavalryman, edited by B. I. Wiley. McCowart-Mercer Press, Jackson, TN, p 205
161. Mosgrove GD (1957) Kentucky Cavaliers in Dixie: reminiscences of a Confederate Cavalryman, edited by B. I. Wiley. McCowart-Mercer Press, Jackson, TN, p 206

162. Mosgrove GD (1957) Kentucky Cavaliers in Dixie: reminiscences of a Confederate Cavalryman, edited by B. I. Wiley. McCowart-Mercer Press, Jackson, TN, p 207
163. Mosgrove GD (1957) Kentucky Cavaliers in Dixie: reminiscences of a Confederate Cavalryman, edited by B. I. Wiley. McCowart-Mercer Press, Jackson, TN, p 208
164. Mosgrove GD (1957) Kentucky Cavaliers in Dixie: reminiscences of a Confederate Cavalryman, edited by B. I. Wiley. McCowart-Mercer Press, Jackson, TN
165. Noe WK (1994) Southwest Virginia's railroad. University of Illinois Press, Urbana, p 115
166. Noe WK (1994) Southwest Virginia's railroad. University of Illinois Press, Urbana, p 111
167. Noe WK (1994) Southwest Virginia's railroad. University of Illinois Press, Urbana, p 129
168. Noe WK (1994) Southwest Virginia's railroad. University of Illinois Press, Urbana, p 112
169. Noe WK (1994) Southwest Virginia's railroad. University of Illinois Press, Urbana
170. Norfolk Southern (2014) http://www.nscorp.com/nscportal/nscorp/Community/NS%20History
171. Norville CR (1991) Tredegar iron works: arsenal of the south. Civil War, September–October 1991, pp 24–26
172. O'Sullivan P (1991) Terrain and tactics. Greenwood Press, West Port, CT, p 121
173. O'Sullivan P (1991) Terrain and tactics. Greenwood Press, West Port, CT
174. Paterson JH (1994) North America. Oxford University Press, New York, p 312
175. Paterson JH (1994) North America. Oxford University Press, New York
176. Pfeil RW, Read JF (1980) Cambrian carbonate platform facies, Shady Dolomite, southwestern Virginia, U. S. A. J Sediment Petrol 50:91–115
177. Pierce EG (1930) A survivor of the Merrimac's crew tells of the Merrimac and Monitor and their famous battle in Hampton Roads. Norfolk and Western Magazine 8(6):384–386
178. Powers J (1981) Confederate niter production. National Speleological Society Bulletin 43:96
179. Powers J (1981) Confederate niter production. National Speleological Society Bulletin 43:94–97
180. Price JL (1994) Coal mining in Montgomery County, Brush Mountain, etc. In: Price JL, Oakley AL, DeHart MC, Jones EE, Prow G, Hodge HK (eds) A Brief history of several coal mines of Montgomery County, VA. Coal Miners Memorial Fund Committee, pp 1–2
181. Proco G (1994) Merrimac mines: a personal history. Southern Printing, Inc., Blacksburg, VA, p 10
182. Proco G (1994) Merrimac mines: a personal history. Southern Printing, Inc., Blacksburg, VA, p 8
183. Proco G (1994) Merrimac mines: a personal history. Southern Printing, Inc., Blacksburg, VA, p 10
184. Proco G (1994) Merrimac mines: a personal history. Southern Printing, Inc., Blacksburg, VA
185. Rachal WME (1953) Salt the South could not savor. Virginia Cavalcade 3:4–7
186. Rains GW (1882) "History of the Confederate Powder Works," An address delivered by invitation before the Confederate survivors' association at its fourth annual meeting, 26 Apr 1882, p 3
187. Rains GW (1882) "History of the Confederate Powder Works," An address delivered by invitation before the Confederate survivors' association at its fourth annual meeting, 26 Apr 1882, p 27–28
188. Rains GW (1882) "History of the Confederate Powder Works," an address delivered by invitation before the Confederate survivors' association at its fourth annual meeting, 26 Apr 1882
189. Ramsey TR (1973) The raid. Kingsport Press Inc., Kingsport, TN
190. Ray CE, Cooper BN, Benninghoff WS (1967) Fossil mammals and pollen in a late Pleistocene deposit at Saltville, Virginia. J Paleontol 41:608–622
191. Roberts JK (1942) Annotated geological bibliography of Virginia. The Dietz Press, Richmond, VA
192. Robertson JI Jr (1993) Lead. In: Current RN (ed) Encyclopedia of the Confederacy. Simon and Schuster, New York, p 913
193. Robertson JI Jr (1993) Cloyds Mountain. In: Current RN (ed) Encyclopedia of the Confederacy. Simon and Schuster, New York, pp 357–358
194. Robertson JI Jr (1997) Stonewall Jackson: the man, the soldier, the legend. Macmillan Publishing, New York, p 348

195. Robertson JI Jr (1997) Stonewall Jackson: the man, the soldier, the legend. Macmillan Publishing, New York
196. Poem by Rudyard Kipling, quoted in Craig JR, Vaughn DJ, Skinner BJ (2001) Resources of the earth; origin, use, and environmental impact. Prentice Hall, Upper Saddle River, NJ, p 52
197. Saltville Historical Foundation (undated) Saltville and the Civil War. Informational Brochure, Saltville
198. Saltville Historical Society (undated) Saltville and salt manufacturing. Informational Brochure, Saltville
199. Sarvis W (1992) The salt trade of nineteenth century Saltville, Virginia. Self-published and handbound in Columbia, Missouri
200. Schmitt MF (ed) (1960) General George Crook: his autobiography. University of Oklahoma Press, Norman, p xv
201. Schmitt MF (ed) (1960) General George Crook: his autobiography. University of Oklahoma Press, Norman, p 301
202. Schmitt MF (ed) (1960) General George Crook: his autobiography. University of Oklahoma Press, Norman
203. Schroeder-Lein GR (1993) Niter and mining bureau. In: Current RN (ed) Encyclopedia of the Confederacy. Simon and Schuster, New York, pp 1146–1148
204. Schroeder-Lein GR (1993) Saltpeter. In: Current RN (ed) Encyclopedia of the Confederacy. Simon and Schuster, New York, p 1363
205. Schult F (1993) Tredegar iron works. In: Current RN (ed) Encyclopedia of the Confederacy. Simon and Schuster, New York, p 1616
206. Schult F (1993) Tredegar iron works. In: Current RN (ed) Encyclopedia of the Confederacy. Simon and Schuster, New York, pp 1616–1617
207. Schultz AP, Bartholomew MJ (2009) Geologic map of the Radford North Quadrangle, Virginia. Virginia Division of Geology and Mineral Resources Open File Report 09-01, 1:24,000-scale geologic map
208. Semple EC (1933) American history and its geographic conditions. Houghton Miflin Company, New York, p 18
209. Semple EC (1933) American history and its geographic conditions. Houghton Miflin Company, New York, p 282
210. Semple EC (1933) American history and its geographic conditions. Houghton Miflin Company, New York, p 290
211. Semple EC (1933) American history and its geographic conditions. Houghton Miflin Company, New York, p 295
212. Semple EC (1933) American history and its geographic conditions. Houghton Miflin Company, New York
213. Sharpe RD (1985) Geology and mining of gypsum in Virginia. In: Glaser JD, Edwards J (eds) Twentieth forum on the geology of industrial minerals. Maryland Geological Survey, Special Publication No. 2, pp 41–49
214. Sipe W (1964) Confederate powder works. Gun Digest, pp 84–89
215. Smith RH (1949) General William Mahone, Frederick J. Kimball and others—a short history of the Norfolk & western railway. The Newcomen Society in North America, New York
216. Smith CH (1981) The land that is Pulaski County. Pulaski County Library Board, Pulaski, VA
217. Smith MO (1987) The identification of Horner's and Heaton's niter works, Bath County, Virginia. National Speleological Society Bulletin 49:21
218. Smith MO (1987) The identification of Horner's and Heaton's niter works, Bath County, Virginia. National Speleological Society Bulletin 49:15–25
219. St. Clair H (1977) Mineral industry in early America. United States Department of the Interior, Bureau of Mines, Washington, DC
220. St. John I (1865) Communication from Secretary of War...Richmond. Confederate States of America, Richmond, 14 Feb 1865
221. Stover JF (1993) Railroads. In: Current RN (ed) Encyclopedia of the Confederacy. Simon and Schuster, New York, p 1298

222. Stover JF (1993) Railroads. In: Current RN (ed) Encyclopedia of the Confederacy. Simon and Schuster, New York, pp 1293–1299
223. Stuart EB (1902) A true story: Saltworks Christmas of 1864. Wytheville Dispatch XLI(22), Tuesday, April 23, 1902, p 3
224. Stuart EB (1902) A true story: Saltworks Christmas of 1864. Wytheville Dispatch XLI(22), Tuesday, April 23, 1902, p 4
225. Stuart EB (1902) A true story: Saltworks Christmas of 1864. Wytheville Dispatch XLI(22), Tuesday, April 23, 1902, p 5
226. Stuart EB (1902) A true story: Saltworks Christmas of 1864. Wytheville Dispatch XLI(22), Tuesday, April 23, 1902
227. Sturgill MH (1990) Abijah Thomas and his octagonal house. Tucker Printing, Marion, VA, p 251
228. Sturgill MH (1990) Abijah Thomas and his octagonal house. Tucker Printing, Marion, VA
229. Sweet PC, Nolde JE (1995) Coal, oil and gas, and industrial and metallic mineral industries in Virginia, 1993. Virginia Division of Mineral Resources, Publication 139
230. Sweet PC, Good RS, Lovett JA, Campbell EVN, Wilkes GP, Meyers LL (1989) Copper, lead, and zinc resources in Virginia. Virginia Division of Mineral Resources Publication 93
231. Tarbuck EJ, Lutgens FK (2005) Earth: an introduction to physical geology, 8th edn. Prentice Hall, Upper Saddle River, NJ
232. United States Geological Survey (2014) http://minerals.usgs.gov/minerals/pubs/commodity/lead
233. United States War Department (1891) The war of the rebellion: a compilation of the official records of the Union and Confederate armies, vol XXXVII, Part 1. Government Printing Office, Washington, DC, p 56
234. United States War Department (1891) The war of the rebellion: a compilation of the official records of the Union and Confederate armies, vol XXXVII, Part 1. Government Printing Office, Washington, DC, p 369
235. United States War Department (1892) The war of the rebellion: a compilation of the official records of the Union and Confederate armies, vol XXXIX, Part 1. Government Printing Office, Washington, DC, p 554
236. United States War Department (1892) The war of the rebellion: a compilation of the official records of the Union and Confederate armies, vol XXXIX, Part 1. Government Printing Office, Washington, DC, p 557
237. Vandiver FE (1952) Ploughshares into swords: Josiah Gorgas and Confederate ordnance. University of Texas Press, Austin, p 199
238. Vandiver FE (1952) Ploughshares into swords: Josiah Gorgas and Confederate ordnance. University of Texas Press, Austin, p 122
239. Vandiver FE (1952) Ploughshares into swords: Josiah Gorgas and Confederate ordnance. University of Texas Press, Austin, p 177
240. Vandiver FE (1952) Ploughshares into swords: Josiah Gorgas and Confederate ordnance. University of Texas Press, Austin, p 197
241. Vandiver FE (1952) Ploughshares into swords: Josiah Gorgas and Confederate ordnance. University of Texas Press, Austin, p 147
242. Vandiver FE (1952) Ploughshares into swords: Josiah Gorgas and Confederate ordnance. University of Texas Press, Austin
243. Virginia Charter (1606) http://avalonlaw.yale.edu/17th_century/va01.asp
244. Virginia Division of Mineral Resources. www.dmme.virginia.gov/DGMR/divisiongeology-mineralresources.shtml
245. Walker GC (1985) The war in Southwest Virginia, 1861–1865. Gurtner Graphics and Printing Co., Roanoke, VA, p 161
246. Walker GC (1985) The war in Southwest Virginia, 1861–1865. Gurtner Graphics and Printing Co., Roanoke, VA
247. Walker GC (1989) Yankee soldiers in Virginia valleys: hunters raid. A&W Enterprises, Roanoke, VA, p 235
248. Walker GC (1989) Yankee soldiers in Virginia valleys: hunters raid. A&W Enterprises, Roanoke, VA

249. Wallace JS (2008) Chemical analysis of firearms, ammunition, and gunshot residue. CRC Press, Taylor and Francis Group, Boca Raton, FL
250. Watson TL (1905) Lead and zinc deposits of Virginia. Virginia Division of Mineral Resources, Bulletin 1
251. Watson TL (1907) Mineral resources of Virginia. J. P. Bell Company, Lynchburg, VA, pp 211–215, 327–335
252. Weaver BW (1962) The mines of Midlothian. Virginia Cavalcade, Winter 1961–1962, p 41
253. Weaver BW (1962) The mines of Midlothian. Virginia Cavalcade, Winter 1961–62, p 43
254. Weaver BW (1962) The mines of Midlothian: Virginia Cavalcade. Winter 1961–1962, pp 40–47
255. Weinberg ET (1980) Austinville mining—1756 to present. Virginia Miner 26(1):11
256. Weinberg ET (1981) Austinville mines close. Virginia Miner 27(4):43
257. Werrell KP (2012) Crook's regulars: the 36th Ohio volunteer infantry regiment in the war of rebellion. Kenneth P. Werrell, Christiansburg, VA
258. Whisonant RC (1996) Geology and the Civil War in southwestern Virginia: the Smyth County salt works, vol 42. Virginia Minerals Virginia, Division of Mineral Resources, Charlottesville, pp 21–30
259. Whisonant RC (1997) Geology and the Civil War in southwestern Virginia: Union Raiders in the New River Valley, vol 43. Virginia Minerals, Virginia Division of Mineral Resources, Charlottesville, pp 29–40
260. Whisonant RC, Watts CW (1993) Neotectonic investigations in the southeastern United States, Part I: Potential seismic triggering of giant bedrock landslides and suspected mass movements at Mountain Lake in the Giles County Seismic Zone. Report prepared for Ebasco Services, Inc., Greensboro, North Carolina
261. Whitman JA (1942) The iron industry of Wythe County. In: Presgraves JM (ed and publisher, 1972) Wythe County Chapters. Southwest Virginia Enterprise, Wytheville, pp 79–115
262. Wikipedia (2014) http://en.wikipedia.org/wiki/Bullet
263. Wikipedia (2014) http://en.wikipedia.org/wiki/History_of_rail_transport
264. Wikipedia (2014) http://en.wikipedia.org/wiki/Lead
265. Wikipedia (2014) http://en.wikipedia.org/wiki/Musket
266. Wikipedia (2014) http://en.wikipedia.org/wiki/Timeline_of_railway_history
267. Wilkes GP (1988) Mining history of the Richmond Coalfield of Virginia. Virginia Division of Mineral Resources Publication 85
268. Williams TH (1965) Hayes of the twenty-third. Knopf, New York
269. Wilson RB (1891) The Dublin raid: grand army of the republic war papers. Paper read before Fred C. Jones post, No. 401, vol 1, Department of Ohio, p 107
270. Wilson RB (1891) The Dublin raid: Grand Army of the Republic War Papers. Paper read before Fred C. Jones post, vol 1, No. 401, Department of Ohio, pp 92–120
271. Wilson G (1932) Smyth County history and traditions. Kingsport Press, Inc., Kingsport, TN
272. Winters HA, Galloway GE Jr, Reynolds WJ, Rhyme DW (1998) Battling the elements: weather and terrain in the conduct of war. Johns Hopkins University Press, Baltimore, MD, p 3
273. Winters HA, Galloway GE Jr, Reynolds WJ, Rhyme DW (1998) Battling the elements: weather and terrain in the conduct of war. Johns Hopkins University Press, Baltimore, MD, p 133
274. Winters HA, Galloway GE Jr, Reynolds WJ, Rhyme DW (1998) Battling the elements: weather and terrain in the conduct of war. Johns Hopkins University Press, Baltimore, MD
275. Wise JS (1899) The end of an era. Houghton Mifflin Company, New York, p 379
276. Wise JS (1899) The end of an era. Houghton Mifflin Company, New York
277. Worsham G (1986) Montgomery historic sites survey, vol 1. Montgomery County, Christiansburg, VA, p 131
278. Worsham G (1986) Montgomery historic sites survey, vol 1. Montgomery County, Christiansburg, VA

279. Yergin D (1991) The prize: the epic quest for oil, money, and power. Simon and Schuster, New York, p 167
280. Yergin D (1991) The prize: the epic quest for oil, money, and power. Simon and Schuster, New York
281. Youngquist WR (1997) Geodestinies: the inevitable control of earth resources over nations and individuals. National Book Company, Portland, OR, p 17
282. Youngquist WR (1997) Geodestinies: the inevitable control of earth resources over nations and individuals. National Book Company, Portland, OR
283. Zabecki DT (2008) Why terrain matters. Mil Hist 25(5):58
284. Zabecki DT (2008) Why terrain matters. Mil Hist 25(5):54–61

Index

Printed in the United States
By Bookmasters